SYMPOSIUM ON INFINITE DIMENSIONAL TOPOLOGY

EDITED BY

R. D. ANDERSON

ANNALS OF MATHEMATICS STUDIES
PRINCETON UNIVERSITY PRESS

Annals of Mathematics Studies

Number 69

SYMPOSIUM ON INFINITE DIMENSIONAL TOPOLOGY

EDITED BY

R. D. ANDERSON

PRINCETON UNIVERSITY PRESS
AND
UNIVERSITY OF TOKYO PRESS

PRINCETON, NEW JERSEY
1972

LC Card: 69-17445
ISBN: 0-691-08087-9
AMS 1968: 4160, 4601, 4610, 4615,
4625, 4630, 4635, 4690, 4785,
5340, 5374, 5720, 5731, 5755

Published in Japan exclusively by
University of Tokyo Press;
in other parts of the world by
Princeton University Press

Printed in the United States of America

PREFACE

The present volume constitutes the Proceedings of the Symposium on Infinite-Dimensional Topology held in Baton Rouge from March 27 through April 1, 1967. The symposium was organized to bring together mathematicians active in research in one or more of the following areas of infinite-dimensional topology: the topology of linear spaces, fixed point theory, differential topology, and pointset topology. In all of these areas, rather striking new results had recently been obtained, and it was believed that a symposium which encouraged the interchange of ideas and information among those active in these various areas should be conducive to further and, hopefully, more broadly applicable research. Recent results of various of the participants have fully justified this belief.

The idea of the symposium grew out of discussions by V. L. Klee, T. Ganea and the undersigned. The organizing committee of the symposium consisted of F. Browder, V. L. Klee, N. Kuiper, R. Palais and R. D. Anderson. The symposium was sponsored by Louisiana State University under funds made available from a National Science Foundation Science Development Grant to Louisiana State University. The sessions were held in the Student Union on the Baton Rouge campus of L.S.U. About 70 mathematicians from a dozen different countries participated. There were about 30 invited hour and half-hour addresses with several problem sessions and many informal discussions. Some of the talks were surveys, some were expository, and others represented current research. The participants were invited to submit papers for the Proceedings. The present volume is the collection of those papers submitted at or shortly after the time of the symposium, and represents contributions in all of the major areas of the symposium. It is hoped that the publication of this volume will stimulate still further research.

The editor sincerely apologizes to the contributors and to other interested mathematicians for the excessive delay in publication of this volume. The delay was his fault and his alone. The editor wishes to thank the Princeton University Press, and particularly John W. Hannon of the Press, for their great help and cooperation in the publication of these Proceedings and for their patience and forbearance with the editor.

R. D. ANDERSON

CONTENTS

SYMPOSIUM ON
INFINITE DIMENSIONAL TOPOLOGY

TOPOLOGICAL EQUIVALENCE OF NON-SEPARABLE REFLEXIVE BANACH SPACES

Ordinal Resolutions of Identity and Monotone Bases*

C. BESSAGA

In 1965 Kadec [5] proved that all separable infinite-dimensional Banach spaces are homeomorphic. The problem, if every non-separable Banach space of density character $\aleph$ is homeomorphic to the Hilbert space $l_2(\aleph)$, is still open. Recently, Troyansky [13] proved that the space $c_0(\aleph)$, of all continuous functions defined on the one-point compactification of a discrete set of cardinality $\aleph$ vanishing at infinity, is homeomorphic to $l_2(\aleph)$. The homeomorphism, he constructed, is a non-separable analogue of the Bernstein-Kadec map [7].

In this paper (Section 2), another (dual) Kadec's map [8] is adopted for the case of (not necessarily separable) Banach spaces having so-called "boundedly complete monotone bases." This makes it possible to show that every reflexive Banach space of any density character $\aleph$ is homeomorphic to the Hilbert space $l_2(\aleph)$. The proof uses the fact that every reflexive space contains a subspace, with the density character equal to that of the whole space, having a boundedly complete monotone basis (Theorem 1 and Proposition 3) and from the result of Bessaga-Pełczynski [2, Theorem 8.2], which states that *every Banach space of density character* $\aleph$ *which contains a subspace homeomorphic to* $l(\aleph)$ *is itself homeomorphic to* $l(\aleph)$ *(and to* $l_2(\aleph)$*).*

The apparatus of monotone bases, generalizing the notion of Schauder bases in separable Banach spaces, is introduced in Section 1. The argu-

* The results of this paper have been published, mostly without proofs, in the paper [3] under the same title.

ments there have some points of similarity with Lindenstrauss [17] and Amir-Lindenstrauss [1].

I would like to express my gratitude to A. Pełczynski for valuable discussion during the preparation of this paper.

Notation. Greek letters (except ξ) denote ordinal numbers. The symbol $\tau_n \to \tau$ means that $\tau_1 \leqq \tau_2 \leqq \cdots$ is a countable sequence and $\lim_n \tau_n = \tau$. X, Y denote real Banach spaces and A, B denote subsets of them. If $A \subset X$ then span A and $\overline{\text{span}}$ A denote the subspace of X spanned on A and the closure of this subspace, respectively. $wX = \inf\{\text{card } A: \overline{\text{span}}\, A = X\}$ is the density character of X. For any subspace B of the conjugate space X^* we denote $B^{\perp} = \{x \in X: f(x) = 0 \text{ for all } f \text{ in } B\}$, the annihilator of B in X. For any ordinal number ξ, by $l(\xi)$ we shall denote the space of all systems $x = \{a_\tau\}_{\tau<\xi}$ of real numbers such that the set $\{\tau: a_\tau \neq 0\}$ is at most countable and $\|x\| = \Sigma_{\tau<\xi}|a_\tau| < \infty$. These spaces are usually denoted by $l(\aleph)$, where $\aleph = \text{card } \xi$.

1. *Projection bases*

DEFINITION 1. Any system of linear projections $\{S_\tau\}_{\tau \leqq \xi}$, $S_\tau: X \to X$, satisfying the following conditions:

(i) $\sup_\tau \|S_\tau\| < \infty$,

(ii) $S_1 x \equiv 0$, $S_\xi x \equiv x$, $S_\alpha X \subset S_\beta X$ for $\alpha \leqq \beta$,

(iii) for any fixed x the function $\phi(\tau) = S_\tau x$ is continuous in τ with respect to the order topology of the closed interval $[1, \xi]$ —will

be called an ordinal resolution of identity* of type ξ in the space X. If moreover: $\dim(S_{\tau+1} - S_\tau)X = 1$ for all τ then the system $\{S_\tau\}$ will be called a projection basis of type ξ; in this case, any system $\{e_\tau; f_\tau\}_{\tau<\xi}$ such that $e_\tau \in X$, $f_\tau \in X^*$ and $f_\tau(x)e_\tau = (S_{\tau+1} - S_\tau)x$, will be called a biorthogonal system associated with the basis $\{S_\tau\}_{\tau \leqq \xi}$. Given a projection basis $\{S_\tau\}$, we shall denote $R_\tau x = x - S_\tau x$ for $\tau \leqq \xi$.

---

* Ordinal resolutions of identity were applied by Lindenstrauss [12] and Amir and Lindenstrauss [1].

Example 1. Let X be a separable Banach space with a Schauder basis $\{x_n\}$ and coefficient-functionals $\{f_n\}$. Then the partial sum projections: $S_n x = \Sigma_{j<n} f_j(x)e_j$ constitute a projection basis in X.

Example 2. $X = l(\xi)$. $S_\tau\{a_\alpha\} = \{b_\alpha\}$, where $b_\alpha = a_\alpha$ for $\alpha < \tau$ and $b_\alpha = 0$ for $\alpha \geqq \tau$, is a projection basis.

PROPOSITION 1. If $\{e_\tau\}_{\tau<\xi}$ is a system of vectors in $X \setminus \{0\}$ with the property that for any $\alpha < \xi$, we have

$$\|x + e_\alpha\| \geqq \|x\| \text{ for all } x \in \overline{\text{span}}\, \{e_\tau\}_{\tau<\alpha} \tag{1}$$

then there exists a projection basis $\{S_\tau\}_{\tau \leqq \xi}$ in the space $\overline{\text{span}}\, \{e_\tau\}_{\tau<\xi}$ such that $e_\tau \in (S_{\tau+1} - S_\tau)x$ for all $\tau < \xi$.

Proof: Let the vectors $\{e_\tau\}_{\tau<\xi}$ satisfy (1); it is obvious that they must then be linearly independent. Let $Z = \text{span}\{e_\tau\}_{\tau<\xi}$. For any $x = \Sigma_{n=1}^{n} t_n e_{\tau_n}$, define $\tilde{S}_\tau x = \Sigma_{\tau_n<\tau} t_n e_{\tau_n}$. $\tilde{S}_\tau$ is a projector of norm 1 and can be (uniquely) extended, by continuity to a projector S_τ defined on $\overline{\text{span}}\, \{e_\tau\}_{\tau<\xi}$. The system of projectors $\{S_\tau\}$ fulfills all the conditions of Definition 1. For instance, the continuity of $S_\tau x$ with respect to τ follows immediately from the fact that functions $g(x, \tau) = S_\tau x$ satisfy the Lipschitz condition with respect to x with the same constant 1 for all τ, and are obviously continuous in τ if x is in the dense set Z.

PROPOSITION 2. If $wX > \aleph_0$ and for every $B \subset X^*$ with card $B < wX$ the set $B^\perp \setminus \{0\}$ is non-empty, then there exists a subspace Y of X with $wY = wX$, having a projection basis of type ξ, the first ordinal of cardinality wX.

Proof: We shall define, by transfinite induction, vectors $\{e_\tau\}_{\tau<\xi}$ satisfying the hypothesis of Proposition 1. Let x_1 be an arbitrary vector in $X \setminus \{0\}$. Assume that we have picked the vectors $\{e_\tau\}_{\tau<\alpha}$ satisfying the condition (1), for some $\alpha < \xi$. Let A_α be the set of all rational linear combinations of vectors $\{e_\tau\}_{\tau<\alpha}$. By Hahn-Banach theorem, for any $x \in A_\alpha$ there is an $f_x \in X^*$, $\|f_x\| = 1$ such that $f_x(x) = \|x\|$. Denote by B_α the collection of such functionals. We have

$$\text{card } B_\alpha \leqq \text{card } A_\alpha \leqq \aleph_0 \circ \text{card } \alpha < wX .$$

Hence, by the assumption of the proposition, there is an $e_\alpha \in B_\alpha \setminus \{0\}$. For every $u \in A_\alpha$ we have $\|u + e_\alpha\| \geqq f_u(u + e_\tau) = \|u\|$, and since A_α is dense in $\overline{\text{span}}\, \{e_\tau\}_{\tau < \alpha}$, we conclude that the vectors $\{e_\tau\}_{\tau \leqq \alpha}$ satisfy also the condition (1).

THEOREM 1 *Every reflexive Banach space* X *contains a reflexive subspace* Y, *with* wY = wX, *having a projection basis.*

Proof: First assume that $wX > \aleph_0$. From the reflexivity of X it follows that $wX = wX^*$. Let $B \subset X^*$, with card $B < wX$. Since $wX > \aleph_0$, we conclude that $w(\overline{\text{span}}\, B) < wX^*$, i.e., span B is a proper subspace of X^*. Hence, by Hahn-Banach theorem, there is an $x \in X^{**} = X$, $x \neq 0$, with $x \in (\overline{\text{span}}\, B)^{\perp} \subset B^{\perp}$. To complete the proof we now apply Proposition 2 and the well-known fact that every subspace of a reflexive space is reflexive.

In the case: $wX = \aleph_0$, the theorem follows from Banach's result stating that every infinite dimensional Banach space contains an infinite dimensional subspace with a Schauder basis (cf. [4]) and from Example 1.

DEFINITION 2. A projection basis $\{S_\tau\}_{\tau \leqq \xi}$ is said to be boundedly complete, provided that for any $\tau_n \nearrow \tau \leqq \xi$ and for any countable sequence of vectors $\{y_k\}$, with the property $S_{\tau_k} y_n = y_k$ for $k \leqq n$ and $\sup_k \|y_k\| < \infty$, the limit $y = \lim_k y_k$ exists.

PROPOSITION 3. Every projection basis in any reflexive Banach space is boundedly complete.

Proof: Let $\{S_\tau\}$ be a projection basis in a reflexive Banach space X; let $\{y_k\}$ be a bounded sequence, with $S_{\tau_k} y_n = y_k$ for $n \geqq k$. Without loss of generality we may additionally assume that $y_1 \neq 0$ and $y_{n+1} \neq y_n$ for all n. It is easy to check that the sequence $x_n = y_{n+1} - y_n$ is a Schauder basis in the space $Y = \overline{\operatorname{span}}\{x_n\}$. Y is reflexive, as a subspace of a reflexive space X. Hence, by a theorem of James [5, Ch. IV, §3, Theorem 3] the basis $\{x_n\}$ is boundedly complete, and therefore $\lim_k y_k = \Sigma_n x_n$ exists, i.e., the original basis $\{S_\tau\}$ is boundedly complete.

PROPOSITION 4. Let X be a Banach space with a projection basis $\{S_\tau\}_{\tau \leqq \xi}$ and let $x \epsilon X$. Then the ordinal number $\alpha = \inf\{\tau : x = S_\tau x\}$ is either isolated or is a limit of a countable sequence of ordinals.

Proof: If α is not isolated, then $\tau_n = \inf\{\tau : \|S_\tau x - x\| \geqq 1/n\} \nearrow \alpha$.

PROPOSITION 5. If X is a Banach space with a projection basis $\{S_\tau\}_{\tau \leqq \xi}$ then there exists in X an equivalent norm $\|\cdot\|$ such that

(v) $\|\cdot\|$ is strictly convex, $\|S_{\tau+1} x\| \geqq \|S_\tau x\|$, $\|R_{\tau+1} x\| \leqq \|S_\tau x\|$

for $x \epsilon X$ and $\tau < \xi$.

Proof: Let $\| \ \|_1$ be the original norm in X. Take a biorthogonal system $\{e_\tau, f_\tau\}_{\tau < \xi}$ associated with the basis, such that $\|e_\tau\|_1 = 1$ for all τ. Then for every $\varepsilon > 0$, $x \epsilon X$ we have $\operatorname{card}\{\tau : |f_\tau(x)| > \varepsilon\} \leqq \aleph$. Let $\tau(n, x)$ be the sequence of all ordinals τ for which $f_\tau(x) \neq 0$ and let $|f_{\tau(1,x)}(x)| \geqq |f_{\tau(2,x)}(x)| \geqq \cdots$. Define

$$\|x\|_2 = \|x\|_1 + \sqrt{\sum_{n=1}^{\infty} |2^{-n} f_{\tau(n,x)}(x)|^2} .$$

$\|\cdot\|_2$ is an equivalent strictly convex norm, cf. Day [6]. Now it is easy to check that the norm

$$\|x\| = \sup_{\alpha > \beta} \|S_\alpha x - S_\beta x\|_2$$

satisfies the condition (v) and is equivalent to the original one.

2. *Topological equivalence of reflexive spaces.*

THEOREM 2. *Let* X *be a reflexive infinite-dimensional Banach space. Then* X *is homeomorphic to the space* $l(\xi)$, *where* ξ *is the first ordinal of cardinality* wX.

According to Theorem 1, Proposition 3 and 5, and the result [2, Theorem 8.2], quoted in the introduction, the proof of this theorem can be reduced to that of the following:

PROPOSITION 6. If X is a Banach space with a boundedly complete projection basis $\{S_\tau\}_{\tau \leqq \xi}$, satisfying the condition (v) of Proposition 5, then there exists a homeomorphism $h\colon X \xrightarrow[\text{onto}]{} l(\xi)$.

Construction of h. Let $\{e_\tau, f_\tau\}_{\tau<\xi}$ be a biorthogonal system associated to the basis and let

$$\bar{b}(x, y) = \|x\| \cdot \|y\| + \sum_{\tau<\xi} (\|S_{\tau+1}y\| - \|S_\tau y\|) \cdot \|R_{\tau+1}x\|, \tag{2}$$

$$\text{for } (x, y) \in X \times X;\ b(x) = \bar{b}(x, x)\,.$$

Then

$$hx = \{(\operatorname{sgn} f(x)) \cdot (b(S_{\tau+1}x) - b(S_\tau x))\}_{\tau<\xi} \in l(\xi)\,. \tag{3}$$

The proof that the map h has desired properties makes use of several lemmas.

LEMMA 1. If X is a Banach space with a monotone basis $\{S_\tau\}_{\tau \leqq \xi}$ fulfilling (v), and the functions $\bar{b}$ and b are defined by (2), then:

(1.1) $\bar{b}(x, y)$ is a norm with respect to each variable, if the other is fixed and $\neq 0$,

(1.2) $\|x\|\,\|y\| \leqq \bar{b}(x, y) \leqq 2\|x\|\,\|y\|$,

(1.3) if $\{x_k\}$ is a (countable sequence of vectors such that

(*) $f_\tau(x_k) \to f_\tau(x_0)$, $b(S_\tau x_k) \to b(S_\tau x_0)$, for $\tau < \xi$ and $b(x_k) \to b(x_0)$,

then $\lim_k \|x_k - x_0\| = 0$.

Proof of (1.1) and (1.2): By (v) we have $\|S_{\tau+1}y\| - \|S_\tau y\| \geqq 0$ and is > 0 only for at most countably many values of τ; moreover

$$\sum_{\tau<\xi} (\|S_{\tau+1}y\| - \|S_\tau y\|) = \|S_\xi y\| = \|y\| .$$

Hence we conclude that for any fixed $y \neq 0$, $\bar{b}(x, y)$ is a well-defined norm in x, and $\|x\|\,\|y\| \leqq \bar{b}(x, y) \leqq \|y\|(\|x\| + \sup_\tau \|R_{\tau+1}x\|) \leqq 2\|x\|\,\|y\|$. Taking into account the continuity of the expressions $S_\tau x$ and $R_\tau x$ with respect to τ and the condition (v), it is easy to check by transfinite induction that

$$\bar{b}(x, y) = \|x\|\,\|y\| + \sum_{2\leqq\alpha<\xi} \|S_\alpha y\|(\|R_\alpha x\| - \|R_{\alpha+1}x\|) .$$

This formula together with (v) gives the conclusion, that for any $x \neq 0$, $\bar{b}(x, y)$ is a norm with respect to y.

Proof of (1.3). In the case where $\text{card}\,\xi < \aleph$, the property (1.3) is obvious. Let $\{x_k\}$ be a sequence of vectors in X; consider the following condition stronger than (*):

(**) $\quad f_\tau x_k \to f_\tau x_0,\ S_\tau x_k \to S_\tau x_0$ for $\tau < \xi$, and $b(x_k) \to b(x_0)$.

LEMMA 2. The conditions (v) and (**) imply the following statements:

(2.1) $\|Rx_0\| \leqq \varliminf \|Rx_k\|$ for $1 \leqq \tau < \xi$,

(2.2) $\bar{b}(x_k, S_\alpha x_k - S_\alpha x_0) \to 0$ for $1 \leqq \alpha < \xi$,

(2.3) $\bar{b}(x_k, x_0) \to \bar{b}(x_0, x_0)$,

(2.4) $\|R_{\tau+1}x_k\| \to \|R_{\tau+1}x_0\|$, whenever $\|S_\tau x_0\| < \|S_{\tau+1}x_0\|$,

(2.5) $x_k \to x_0$.

Of course, establishing (2.5) gives an inductive proof of (1.3).

Proof of (2.1). (Cf. Kadec [10], [11], Pełczynski [14]). Let $R_\tau^\alpha x = S_\alpha x - S_\tau x$, $\tau \leqq \alpha$. By (v), $\{R_\tau^\alpha x\}$ is non-decreasing with respect to α, and, by (**), $R_\tau^\alpha x_k \to R_\tau^\alpha x_0$, for $\tau \leqq \alpha < \xi$. Let $\varepsilon > 0$; according to the definition of a monotone basis, there exists an $\alpha(\varepsilon)$ such that $\|R_\tau x_0\| \leqq \|R x_0\| + \varepsilon$ for $\alpha > \alpha(\varepsilon)$. On the other hand, for any $\alpha < \xi$ and $e > 0$ there is an $k(\alpha, e)$ with $\|R_\tau^\alpha x_0\| \leqq \|R_\tau^\alpha x_k\| + \varepsilon \leqq \|R_\tau x_k\| + \varepsilon$ for $k > k(\alpha, \varepsilon)$. Hence $\|R_\tau x_0\| \leqq \|R_\tau x_k\| + 2\varepsilon$ for $k > k(\alpha, \varepsilon)$, which gives (2.1).

Proof of (2.2). By (1.2), $\sup \|x_n\| = M < \infty$, and $0 \leqq \bar{b}(x_k, S_\alpha x_k - S_\alpha x_0) \leqq 2M\|S_\alpha x_0\|^2 \to 0$.

Proof of (2.3). Let $\varepsilon > 0$. Take an $\alpha < \xi$ with $b(x_0) \leqq \bar{b}(x_0, S_\alpha x_0) + \varepsilon$. By (1.1), (1.2), (2.1), (2.2), $b(x_0) \leqq \varepsilon + \bar{b}(x_0, S_\alpha x_0) \leqq \varepsilon + \underline{\lim}_k\ \bar{b}(x_k, S_\alpha x_0)$ $\leqq \varepsilon + \underline{\lim}_k\ (\bar{b}(x_k, S_\alpha x_k) + \bar{b}(x_k, S_\alpha x_k - S_\alpha x_0)) = \varepsilon + \underline{\lim}_k\ \bar{b}(x_k, S_\alpha x_k) + 0 \leqq \varepsilon + \underline{\lim}_k\ \bar{b}(x_k, x_k) = \varepsilon + b(x_0)$. Since $\varepsilon > 0$ was arbitrarily chosen, we obtain (2.3)

Proof of (2.4). Suppose $\|S_\alpha x\| < \|S_{\alpha+1} x\|$ and $\|R_{\alpha+1} x_k\| \not\to \|R_{\alpha+1} x_0\|$, i.e., by (2.1), $\underline{\lim}_k\ \|R_{\alpha+1} x_k\| > \|R_{\alpha+1} x_0\|$. Then we would get

$$\underline{\lim}_k\ \bar{b}(x_k, x_0) \geqq \underline{\lim}_k\ \|R_1 x_k\|\ \|x_0\| + \underline{\lim}_k \sum_{\tau \neq \alpha} (\|S_{\tau+1} x_0\| - \|S_\tau x_0\|)\|R_{\tau+1} x_k\|$$

$$+ \underline{\lim}_k\ (\|S_{\alpha+1} x_0\| - \|S_\alpha x_0\|)\,\|R_{\alpha+1} x_k\| > b(x_0)\,,$$

which contradicts (2.3).

Proof of (2.5). Let $\alpha = \inf\{\tau : S_\tau x_0 = x_0\}$. We shall consider separately three cases:

The case where α is a limit ordinal. By the condition (v) and Proposition 4, we conclude that there are $\tau_n \to \alpha$ such that $\|S_{\tau_n} x_0\| < \|S_{\tau_n+1} x_0\|$. Let $\varepsilon > 0$. Take a $\beta = \tau_{n_0} + 1$, with $\|R_\beta x_0\| < \varepsilon/4$. By (4) and (**), there exists a $k(\varepsilon)$ such that

$$|\,\|R_\beta x_k\| - \|R_\beta x_0\|\,| < \varepsilon/4, \quad \|S_\beta x_k - S_\beta x_0\| < \varepsilon/4$$

for $k > k(\varepsilon)$. Thus for $k > k(\varepsilon)$, we have

$$\|x_k - x_0\| \leqq \|S_\beta x_k - S_\beta x_0\| + \|R_\beta x_k\| + \|R_\beta x_0\| < \varepsilon/4 + \varepsilon/2 + \varepsilon/4 = \varepsilon.$$

The case $\alpha = \gamma + 1 < \xi$. By (**) $S_\alpha x_k \to S_\alpha x_0 = x_0$. Since $\|S_\gamma x_0\| < \|S_{\gamma+1} x_0\|$, the statement (4) gives: $\|R_\alpha x_k\| \to \|R_\alpha x_0\| = 0$. Hence $x_k = S_\alpha x_k + R_\alpha x_k \to x_0$.

The case $\xi = \gamma + 1$. By (**), we have $x_k = S_\xi x_k = S_\gamma x_k + f_\gamma(x_k) e_\gamma \to S_\gamma x_0 + f_\gamma(x_0) e_\gamma = x_0$.

Proof of Proposition 6. In the case of the space $l(\xi)$ by $\{S_\tau\}$ we shall denote the natural basis of Example 2. $\{f_\tau\}_{\tau<\xi}$ will be the coordinate functionals, i.e., $f_\tau(\{a_\alpha\}) = a_\tau$. Let us start with the following:

LEMMA 3. We have:

(3.1) $\|S_\tau hx\| = b(S_\tau x)$ for $\tau \leqq \xi$, $\|x\|^2 \leqq \|hx\| = b(x) \leqq 2\|x\|^2$.

(3.2) For any sequence $\{x_k\}$ in X the following conditions are equivalent: (a) $x_k \to x_0$,

(b) $f_\tau(x_k) \to f_\tau(x_0)$, $b(S_\tau x_k) \to b(S_\tau x_0)$ for $\tau < \xi$ and $b(x_k) \to b(x_0)$,

(c) $f_\tau(hx_k) \to f_\tau(hx_0)$, $\|S_\tau hx_k\| \to \|S_\tau hx_0\|$ for $\tau < \xi$ and $\|hx_k\| \to \|hx_0\|$,

(d) $hx_k \to hx_0$.

(3.3) h is a homeomorphism into $l(\xi)$.

Proof: The statement (3.1) and implications: (b) $\Longrightarrow$ (a), (b) $\Longrightarrow$ (c) follow from the definition of the map h and from the Lemma 1. From the continuity of S_τ, f_τ and b follows that (a) implies (b). The equivalence of the conditions (c) and (d) is an obvious property of the space $l(\xi)$. It remains to prove that (c) implies (b).

Assuming (c), from the statement (3.1) we obtain

(4) $$b(S_\tau x_k) \to b(S_\tau x_0) \text{ for all } \tau \leqq \xi.$$

Now suppose that

(5) $$f_\tau(x_k) \to f(x_0) \text{ for } \tau < \alpha ,$$

and denote $f_\alpha(x_k) = t_k$. By (4), $b(S_\alpha x_k + t_k e_\alpha) \to b(S_{\alpha+1} x_0)$, and using (5), (1.2) and (3.1) we conclude

(6) $$b(S_\alpha x_0 + t_k e_\alpha) \to b(S_{\alpha+1} x_0) .$$

From the definition of h and the condition (c) we obtain

(7) $$\operatorname{sgn} t_k \to \operatorname{sgn} f_\alpha(hx_0) .$$

From the formula (2) defining the function b, the condition (v) of Proposition 5 and Lemma 1 it follows that the function $g(t) = b(S_\alpha x_0 + te)$ is strictly increasing for $t > 0$, strictly decreasing for $t < 0$ and continuous. Hence, by (6) and (7) we obtain

$$f_\alpha(x_k) = t_k \to f_\alpha(x_0) .$$

Since, the inductive hypothesis (5) is obviously true for $\alpha = 1$, by induction we establish the condition (b).

The statement (3.3) is an obvious consequence of (3.2).

Proof that h *is onto.* For any $z \in l(\xi)$ denote $\sigma(z) = \inf\{\tau: S_\tau z = z\}$. If $\sigma(z) = 1$, then $z = 0 = h0$. To complete the inductive proof we have to show:

(***) For any $z \in l(\xi)$ the condition $\{y \in l(\xi): \sigma(y) < \sigma(z)\} \subset hX$ implies $z \in hX$.

The case $\sigma(z) = \gamma + 1$. Take a vector x in X with $hx = S_\gamma z$. Of course $\varepsilon = \operatorname{sgn} f_\gamma(z) \neq 0$. As we have observed, the function $g(t) = b(x + \varepsilon t e_\gamma)$ is continuous and increasing for $t > 0$. Hence there is a t_0, with $g(t_0) = b(z)$. It is easily seen that $z = h(x + \varepsilon t_0 e_\gamma)$.

The case where $\sigma(z)$ *is a limit ordinal.* By Proposition 4, there are $\tau_n \to \sigma(z)$ such that $S_{\tau_n} z \to z$. Take $x_n \in X$ with $hx_n = S_{\tau_n} z$. It is clear that the sequence $\{x_n\}$ has the property stated in the Definition 2, and since the basis $\{S_\tau\}$ is boundedly complete, we conclude that $x = \lim_n x_n$ exists and that $z = hx$.

REMARK 2. In the case where the norm $\|\cdot\|$ of the space X is uniformly convex, we may replace the functional $b(x)$ by the norm $\|x\|$.

3. *Problems*

Problem 1. Is it true that every conjugate Banach space is homeomorphic to a space $l(\xi)$?

REMARK 3. For the spaces $X = Y^*$ for which $wX = wY$, the answer is "yes."

Problem 2. Is it true that every Banach space with a monotone basis is homeomorphic to a space $l(\xi)$? (Cf. Kadec [9].)

Problem 3. Is it true that every Banach space with an unconditional basis is homeomorphic to a space $l(\xi)$?

The affirmative answer to this problem would follow from:

CONJECTURE 1. Assume that X is a Banach space with an unconditional basis of cardinality $\aleph \geqq \aleph_0$ and X does not contain any subspace isomorphic to $c_0(\aleph)$. Then there is a subspace Y of X such that no subspace of Y is isomorphic to c_0, and $wY = \aleph$.

Problem 4. Is it true that every space $C(Q)$, of all continuous functions on a compact space Q, is homeomorphic with a space $l(\xi)$?

UNIVERSITY OF WARSAW, WARSAW, POLAND
and LOUISIANA STATE UNIVERSITY, BATON ROUGE, LA.

REFERENCES

[1] D. Amir and J. Lindenstrauss, "On topological structure of weakly compact sets in Banach spaces," to appear.

[2] C. Bessaga, "On topological classification of complete linear metric spaces," *Fund. Math., 56* (1965), 251-288.

[3] _____, "Topological equivalence of non-separable reflexive Banach spaces. Ordinal resolutions of identity and monotone bases," *Bull. Acad. Polon. Sci., Ser. sci. math. astr. et phys.*, (1967),

[4] _____, and A. Pełczynski, "On bases and unconditional convergence of series in Banach spaces," *Studia Math., 17* (1958), 151-164.

[5] M. M. Day, *Normed Linear Spaces*, Berlin-Göttingen-Heidelberg, 1962.

[6] _____, "Strict convexity and smoothness of normed spaces," *Trans. Amer. Math. Soc., 78* (1955), 516-528.

[7] M. I. Kadec, "On homeomorphism of certain Banach spaces" (Russian), *Dokl. Akad. Nauk. SSSR, 163* (1953), 465-468.

[8] _____, "On strong and weak convergence" (Russian), *ibidem, 122* (1958), 23-25.

[9] _____, "On topological equivalence of all separable Banach spaces" (Russian), *ibidem, 163* (1966), 23-25.

[10] _____, "On spaces isomorphic with locally uniformly convex spaces," (Russian) *Izvestia Vuzov, Matematika, 6 (13)* (1959), 51-57.

[11] _____, Letter to the editor (Russian), *ibidem 6* (1961), 139-141.

[12] J. Lindenstrauss, "On reflexive spaces having the metric approximation property," *Israel. Journ. Math., 3* (1965), 199-204.

[13] S. Troyansky, "On topological equivalence of $c_0(\aleph)$ and $1(\aleph)$," (Russian), to appear.

[14] A. Pelczynski, "On Kadec's paper: 'Topological equivalence of all separable Banach space,'" Mimeographed report delivered in seminar on infinite-dimensional topology, University of Washington, 1966.

ON TOPOLOGICAL CLASSIFICATION OF

NON-SEPARABLE BANACH SPACES

C. BESSAGA AND M. I. KADEC

The problem of topological classification of separable Banach spaces has been completely solved; any such a space, if infinite-dimensional, is homeomorphic to the Hilbert space $l_2(\aleph_0)$ [17]. The problem, if every (non-separable) Banach space is homeomorphic to a Hilbert space, is still open. However in a few interesting cases (for instance for reflexive spaces [5]) the affirmative answer has been established. This supports the conjecture that every Banach space is homeomorphic to a Hilbert space, i.e., that the only topological invariant of Banach spaces is their density character.

Most of the facts on topological equivalence of Banach spaces have been obtained by combining so-called "coordinate methods" and "decomposition methods" with some result of linear character. These methods have found also some applications in case of linear metric spaces which are not Banach spaces: [4], [20], see also [6], [7].

The sections 1 and 2 are devoted to describing coordinate methods in an ordinal number set-up (generalization of "separable" coordinate methods: [3], [13] – [16], [19], [20]) and to outlining their applications. In section 3 we state main theorems of decomposition type and list the results which can be obtained by help of these theorems.

Notation. The letters: α, β, γ, τ, ν denote ordinal numbers; the first ordinal number of a cardinality $\aleph$ will be also denoted by $\aleph$. e, ξ, t denote real numbers. x, y, z denote vectors. All the Banach spaces considered here are over the real scalars. $c_0(\nu)$ is the Banach space of all $x = \{\xi_\alpha\}_{\alpha<\nu}$ such that the set $\{\alpha: |\xi_\alpha| > \varepsilon\}$ is finite for every $\varepsilon > 0$, under the norm $\|x\| = \sup_\alpha |\xi_\alpha|$. $l_p(\nu)$, $p \geq 1$, is the space of all $x = \{\xi_\alpha\}_{\alpha<\nu}$ such that the set $\{\alpha: \xi_\alpha \neq 0\}$ is at most countable and $\|x\| = (\Sigma_\alpha |\xi_\alpha|^p)^{1/p}$. In the sequel, the space $l_1(\nu)$, denoted briefly $l(\nu)$, is in a sense a "test space"; of course it could be replaced by the Hilbert space $l_2(\nu)$, because the latter space is homeomorphic to the first under the map $\{\xi_\alpha\} \to \{\xi_\alpha|\xi_\alpha|\}$, see also Theorem 1.

The symbol "$\simeq$" denotes the relation of being homeomorphic.

1. *Bernstein maps:* B-*systems and* co-B-*systems*

Let X be a Banach space, and let b be a δ-*modular* on X, i.e., a continuous non-negative functional defined on X such that: $\lim_n b(x_n) = 0$ iff $\lim_n x_n = 0$ and b(tx) is non-increasing in t for every $x \in X$, $t \geq 0$. A closed subspace L of X is called a b-*Čebyšev subspace* provided that for any x in X there is the unique b-*nearest point* $Px \in L$, i.e., Px has the property: $z = Px$ iff $z \in L$ and $b(x - z) = \inf\{b(x-y): y \in L\}$.

Any system $<b, \{L_\alpha\}_{\alpha\leq\nu}>$, where b is a δ-modular on X and L_α are b-Čebyšev subspaces of X, will be called a *generalized* [co-] T-*system* iff $L_1 = \{0\}$, $L_\nu = X$, $L_\beta \subset L_\alpha$ for $\alpha > \beta$, $L_\alpha = \overline{\cup_{\beta<\alpha} L_{\beta+1}}$ for $\alpha \leq \nu$ $\dim(L_{\alpha+1}/L_\alpha) = 1$ for $\alpha < \nu$ [$L_1 = X$, $L_\nu = \{0\}$, $L_\beta \supset L_\alpha$ for $\alpha > \beta$, $L_\alpha = \cap_{\beta<\alpha} L_{\beta+1}$, $\dim(L_\alpha/L_{\alpha+1}) = 1$].

An *oriented* generalized [co-] T-system

$$(*) \qquad <b, \{L_\alpha\}_{\alpha\leq\nu}, \{f_\alpha\}_{\alpha<\nu}>$$

is a generalized [co-] T-system together with a family of linear functionals $f_\alpha \in L^*_{\alpha+1}$ [$f_\alpha \in L^*_\alpha$] such that $f_\alpha(x) = 0$ iff $x \in L_\alpha$[$x \in L_{\alpha+1}$].

Let P: $X \to L_\alpha$ be the b-nearest-point map and let $d_\alpha(x) = b(x - P_\alpha x)$, $\varepsilon_\alpha(x) = \operatorname{sgn} f_\alpha(P_\alpha x)$[$\varepsilon_\alpha(x) = -\operatorname{sgn} f_\alpha(P_{\alpha+1} x)$]. The system (*) induces the Bernstein map h: $X \to l(\nu)$, where $hx = (d_\alpha(x) - d_{\alpha+1}(x)) \cdot \varepsilon_\alpha(x)$.

An oriented generalized [co-] T-system will be called a *generalized [co-] B-system,* provided that h is a homeomorphism onto $l(\nu)$. Omitting the adjective "generalized" will indicate that $b(x) = \|x\|$, the norm of the space.*

It follows directly from the definition of generalized [co-] B-systems, that in order to establish a homeomorphism between a Banach space X and a space $l(\nu)$ it suffices to find in X either a generalized B-system or a generalized co-B-system. To check if a given generalized [co-] T-system is a generalized [co-] B-system is, in general, not difficult. For instance, in the countable case (more precisely if $\nu = \aleph_0$) every oriented T-system the Bernstein map of which is one-to-one, is a B-system (because all oriented T-systems, with $\nu = \aleph_0$, have the property that h is continuous, onto $l(\aleph_0)$ and h^{-1} preserves precompactness).

Example A. Let X be either $c_0(\nu)$ or $l_p(\nu)$, $p \geq 1$. Let $L_\alpha = \{\{\xi_\tau\} \in X: \xi_\alpha = 0 \text{ for } \tau \geq \alpha\}$, $L^\alpha = \{\{\xi_\tau\} \in X: \xi_\tau = 0 \text{ for } \tau < \alpha\}$, $f_\alpha(\{\xi_\tau\}) = \xi_\alpha$. Then:

1) $<\|\cdot\|, \{L_\alpha\}, \{f_\alpha\}>$ [$<\|\cdot\|, \{L^\alpha\}, \{f_\alpha\}>$] is a [co-] B-system in each $l_p(\nu)$.

2) There exists an equivalent norm $|||\cdot|||$ in $c_0(\nu)$ such that $<|||\cdot|||, \{L_\alpha\}, \{f_\alpha\}>$ is a B-system in $c_0(\nu)$ – Troyanski [24], cf. [13], [10]. This norm can be defined by Day's [11] formula: $|||x||| = \|x\| + \Sigma_n 2^{-n} |\xi_{p_n}|$, where $\{\xi_{p_n}\}$ is the sequence of all non-zero coordinates of x ordered in their non-increasing way.

Example B. Let X be a separable conjugate Banach space, $X = Y^*$. Let $\{y_n\}$ be a linearly independent sequence in Y such that $x(y_n) = 0$ for all n implies $x = 0$ and $\lim_k x_k(y_n) = x(y_n)$ for $n = 1, 2, \ldots$, $\lim_k \|x_k\| = \|x\|$ imply $\lim_k \|x_k - x\| = 0$. Denote $L^n = \{x: x(y_i) = 0 \text{ for } i < n\}$, $f_n(x) = x(y_n)$. Then $<\|\cdot\|, \{L^n\}, \{f_n\}>$ is a co-B-system. Using this argument it is possible to show that every separable conjugate Banach space admits general-

* The term "T-system" has been introduced by Klee and Long [20] (T for Tchebysheff). B-systems were called there "T-systems with Bernstein property."

ized co-B-systems, and therefore is homeomorphic to $1(\aleph_0)$; see [15], [16], [20].

By Example A we obviously have

THEOREM 1. *The spaces* $c_0(\nu)$ *and* $1_p(\nu)$, $p \geq 1$, *are homeomorphic.*

2. *B and co-B-systems related to projection bases*

Let X be a Banach space with a projection basis $\{S_\tau\}$. The symbols R_τ, e_τ, f_τ will have the same meaning as in [5]. We shall consider two systems of subspaces of X, $L_\alpha = S_\alpha X$ and $L^\alpha = R_\alpha X$.

Example C. X is either $c_0(\nu)$ or $1_p(\nu)$. $S_\alpha\{\xi_\tau\} = \{\xi'_\tau\}$, where

$$\xi'_\tau = \begin{cases} \xi_\tau & \text{for } \tau < \alpha \\ 0 & \text{for } \tau \geqq \alpha . \end{cases}$$

The subspaces L_α and L^α are now the same as in Example A.

A projection basis $\{S_\alpha\}_{\alpha \leq \nu}$ is said to be:

(a) *orthogonal,* iff L_α and L^α are norm-Čebyšev and S_α, R^α are the norm-nearest point maps onto L_α and L^α, respectively,

(b) *boundedly complete,* iff for any $\alpha_1 < \alpha_2 < \alpha_3 < \dots$ and for any sequence $\{x_n\}$, with $x_n = (S_{\alpha_n} - S_{\alpha_{n+1}})x_n$ the condition $\sup_n \|\Sigma_{i=1}^n x_i\| < \infty$ implies the convergence of $\Sigma_n x_n$.

(c) *unconditional,* iff there exists a σ-additive projection-valued measure $E(\cdot)$ defined on the σ-field of all subsets of the segment $[1, \nu]$ such that $E\{\alpha: \alpha < \tau\} = S_\tau$, for all $\tau \leq \nu$.

PROPOSITION 1. If X has a projection basis, then it is possible to renorm X in such a way that the basis becomes orthogonal. (Cf. [5, Proposition 5].)

PROPOSITION 2. Every projection basis in a reflexive Banach space is boundedly complete.

PROPOSITION 3. Any reflexive Banach space contains a (reflexive) subspace of density character equal to that of the whole space, admitting a projection basis: [5].

THEOREM 2 *If* X *has a boundedly complete orthogonal projection basis* $\{S_\alpha\}$, f_α, R_α *and* L^α *are defined as above, and* $b(x) = \|x\|^2 + \Sigma_\alpha (\|S_{\alpha+1}x\| - \|S_\alpha x\|)\|R_{\alpha+1}x\|$, *then* $<b, \{L^\alpha\}, \{f_\alpha\}>$ *is a generalized co-B-system in* X. *Hence every Banach space with a boundedly complete projective basis of type* ν *is homeomorphic to* $l(\nu)$, see [5].

REMARK 1. In the case where the space X is uniformly convex, the above b(x) can be replaced by the norm of the space. S. Troyanski has shown that also in the case of boundedly complete unconditional bases one can use a norm in place of b(x).

Problem 1. Does every Banach space admit a generalized B-system?

Problem 2. Does every conjugate Banach space admit a generalized co-B-system?

REMARK 2. If X is separable and there exist closed bounded convex subsets of X without extreme points (for instance $X = L(0, 1)$), then X does not possess any co-B-system, see [8].

Problem 3. Let X be a Banach space with a projection basis $\{S_\alpha\}$. Does there exist a δ-modular b such that $<b, \{L_\alpha\}, \{f_\alpha\}>$ [$<b, \{L^\alpha\}, \{f_\alpha\}>$] is a generalized [co-] B-system?

The problem is open even in the separable case; we have however:

THEOREM 3. *If* $\{S_\alpha\}_{\alpha \leq \nu}$ *is a projection basis in a Banach space* X, *then there exist an equivalent norm* $|||\cdot|||$ *and a* δ*-modular* b *such that* $<b, \{L^\alpha\}, \{f_\alpha\}>$ *is a co-T* system the Bernstein map of which restricted to the sphere* $\{x \in X: |||x||| = 1\}$ *is a homeomorphism onto the unit sphere of* $l(\nu)$.

(Take the functional F(x) defined in [18] and set

$$b(x) = \begin{cases} 1 - F(x) & \text{for } |||x||| \leq 1 \\ |||x||| & \text{for } |||x||| > 1 \end{cases}).$$

Problem 4. Let X be a Banach space with a basis $\{S_\alpha\}_{\alpha \leq \nu}$. Does there exist a new equivalent norm and a δ-modular b such that

$\langle b,\{L^{\alpha}\},\{f_{\alpha}\}\rangle$ is a co-T.system and its Bernstein map restricted to the new unit sphere of X is a homeomorphism onto the unit sphere of $l(\nu)$?

More general:

Problem 5. Let X be a Banach space. Does there exist in X an oriented generalized [co-] T-system $\langle b,\{L_{\alpha}\},\{f_{\alpha}\}\rangle$ such that $U = \{x : b(x) \leq 1\}$ is bounded and convex and the Bernstein map restricted to ∂U is a homeomorphism onto the unit sphere of $l(\nu)$?

The affirmative answer to this problem would imply $X \simeq l(\nu)$.

3. *Decomposition theorems; applications*

Let X and Y be topological spaces. We shall write $Y|X$ provided that there exists a space W such that $X \simeq Y \times W$.

The next two theorems concern Fréchet spaces (i.e., locally convex complete linear metric spaces) and in particular are valid for Banach spaces:

THEOREM 4. *Let X and Y be Fréchet spaces. If either Y is a subspace of X or there is a continuous linear map* $T: X \xrightarrow[\text{onto}]{} Y$, *then* $Y|X$.

This is an easy corollary from Michael's Fréchet space version of Bartle-Graves result, see [22] and [2].

THEOREM 5. *Let X be a Fréchet space of density character $\aleph$. If $l(\aleph)|X$ then $X \simeq l(\aleph)$* [6, Th. 8.2].

COROLLARY 1. $(l(\aleph))^{\aleph_0} \simeq l(\aleph)$.

From Propositions 3, 1, 2 and Theorems 2, 4, 5 it follows:

THEOREM 6. *Every reflexive Banach space is homeomorphic to a space $l(\nu)$, see* [5].

COROLLARY 2. *If X is a Banach space of density character $\aleph$ such that either X or X^* contains a reflexive subspace of density character $\aleph$, then $X \simeq l(\aleph)$* ([6], 9.3. xix).

The next theorem summarizes the known results on topological equivalence of spaces C(Q), of all continuous functions on the compact space Q, under the sup-norm:

THEOREM 7. *Let* Q *be a compact Hausdorff space. Each of the conditions (1)-(4) listed below is sufficient in order that* C(Q) *be homeomorphic with a space* $l(\nu)$*:*

(1) Q *is the one-point compactification of a discrete space,*

(2) Q *is the Stone-Cech compactification of a discrete space,*

(3) Q *is a topological group,*

(4) Q *contains a closed subset* D *such that* $C(D) \simeq l(\aleph)$, *where* $\aleph$ *is the density character of the space* C(Q).

REMARK 3. The condition (3) can be replaced by a weaker one:

(3′) Q admits a countable sequence of Baire measures μ_n such that the measure algebras $B(\mu_n, Q)$ are homogeneous in the sense of Maharam [21] and the supremum of density characters of these algebras is equal to the density character of the space C(Q).

The sufficiency of (1) follows from the Troyanski's result—Theorem 1 of this paper. The other condition has been established by Pełczynski [23], see also [6].

Problem 6. Let X be a C(Q) space with density character $\aleph \geq \aleph_0$. Must X be homeomorphic to the space $l(\aleph)$?

THEOREM 8. *Every abstract* L*-space is homeomorphic to a space* $l(\nu)$ ([6], 9.3 xx).

The above facts seem to suggest that the solution of the general classification problem of (non-separable) Banach spaces can be perhaps achieved by studying:

1) geometrical properties of Banach spaces connected with the existence of "nice" norms and δ modulars and "nice" generalized [co-] T-systems,

2) some isomorphic properties of Banach spaces, mainly the structure of subspaces and linear images of a given space.

We may expect also that the investigation of structural properties of Fréchet spaces will allow to reduce the classification problem of Fréchet spaces to that of Banach spaces. (In the separable case this was possible thanks to Anderson's theorem [1]: $l(\aleph_0) \simeq s$, the countable product of lines, Eidelheit's result [12] stating that every non-normable Fréchet space can be linearly mapped onto s, and Theorem 4.5.)

REFERENCES

[1] R. D. Anderson, "Hilbert space is homeomorphic to the countable infinite product of lines," *Bull. Amer. Math. Soc.*, *72* (1966), 515-519.

[2] R. G. Bartle and L. M. Graves, "Mappings between function spaces," *Trans. Amer. Math. Soc.*, *72* (1952), 400-413.

[3] S. Bernstein, "Sur le probléme inverse de la théorie de meilleure approximation des fonctions continues," *C. R. Acad. Sci. Paris*, *206* (1938), 1520-1523.

[4] C. Bessaga, "Some remarks on homeomorphisms of $\aleph_0$-dimensional linear metric spaces," *Bull. Acad. Pol. Sci.*, ser. sci., math., astr. et phys., *11* (1963), 159-163.

[5] _____, "Topological equivalence of non-separable reflexive Banach spaces. Ordinal resolutions of identity and monotone bases," this issue p. 3.

[6] _____, "On topological classification of complete linear metric spaces," *Fund. Math.*, *56* (1965), 251-288.

[7] _____ and A. Pełczynski, "Some remarks on homeomorphisms of F-spaces," *Bull. Acad. Pol. Sci.*, ser. sci., math., astr. et phys., *10* (1962), 265-270.

[8] _____, "On extreme points in separable conjugate spaces," *Israel J. Math.*, *4* (1966), 262-264.

[9] H. Corson and V. Klee, "Topological classification of convex sets," *Proc. Symp. Pure Math., 7,* "Convexity," *Amer. Math. Soc., 37-51.*

[10] R. O. Davis, "A norm satisfying the Bernstein condition," *Studia Math., 29* (1967), 219-220.

[11] M. M. Day, "Strict convexity and smoothness of normed spaces," *Trans. Amer. Math. Soc., 70* (1955), 516-528.

[12] M. Eidelheit, "Zur Theorie der Systeme linearen Gleichungen," *Studia Math., 6* (1936), 139-148.

[13] M. I. Kadec, "On homeomorphism of certain Banach spaces (Russian)," *Dokl. Akad. Nauk SSSR* (N. S.) *92* (1953), 465-468.

[14] ———, "On topological equivalence of uniformly convex spaces (Russian)," *Usp. Mat. Nauk, 10* (1955), 137-141.

[15] ———, "On connection between weak and strong convergence (Ukrainian)," *Dopovidi Akad. Nauk Ukrain., RSR 9* (1959), 465-468.

[16] ———, "On strong and weak convergence (Russian)," *Dokl. Akad. Nauk SSSR* (N. S) *122* (1958), 13-16.

[17] ———, "On topological equivalence of cones in Banach spaces (Russian)," *ibidem, 162* (1965), 1241-1244.

[18] ———, "On topological equivalence of all separable Banach spaces (Russian)," *ibidem, 167* (1966), 23-25.

[19] V. Klee, "Mappings into normed spaces," *Fund. Math., 49* (1960), 25-34.

[20] ——— and R. G. Long, "On a method of mapping due to Kadec and Bernstein," *Archiv. der Math., 8* (1957), 280-285.

[21] D. Maharam, "On homogeneous measure algebras," *Proc. Nat. Acad. Sci., USA, 28* (1942), 108-111.

[22] E. Michael, "Convex structure and continuous selections," *Canad. J. Math., 11* (1959), 556-576.

[23] A. Pełczynski, "Linear extensions and averagings of continuous functions," *Rozprawy Matematyczne*, in print.

[24] S. Troyanski, "On topological equivalence of $c_0(\aleph)$ and $l(\aleph)$ (Russian," *Bull. Acad. Polon. Sci.*, ser. sci., math., astr. et phys. (1967) 389-396.

Warsaw and Kharkov

ON HOMOTOPY PROPERTIES OF COMPACT SUBSETS OF THE HILBERT SPACE

KAREL BORSUK

One understands usually by *homotopy properties* such topological properties of a space which depend only on the homotopy type (in the sense of Hurewicz [3]). However, the notion of the homotopy type is adapted rather to spaces with a regular local structure (as polyhedra, or more generally, ANR-spaces). In the case of spaces with a more complicated local structure, the classical notion of the homotopy type does not give much information about the global structure of the space.

The aim of these notes is to introduce some notions, which in the case of compact ANR-spaces coincide with the classical notions of the theory of homotopy, but show a more intimate connection with the global topological properties of space than the classical notions.

§1. FUNDAMENTAL SEQUENCES AND FUNDAMENTAL CLASSES.

Let X and Y be two compact sets lying in the Hilbert space H. A sequence $\{f_n\}$ of maps (i.e., of continuous functions) $f_n\colon H \to H$ will be said to be a *fundamental sequence from* X *to* Y provided for every neighborhood V of Y (in the space H) there exists a neighborhood U of X (in H) such that the restrictions f_n/U, f_{n+1}/U are homotopic in V for almost all n, that is, there exists a homotopy $\phi\colon U \times \langle 0, 1\rangle \to V$ such that $\phi(x, 0) = f_n(x)$, $\phi(x, 1) = f_{n+1}(x)$ for every point $x \in U$. We write then $f_n/U \simeq f_{n+1}/U$ in V.

Two fundamental sequences $\underline{f} = \{f_n\}$, $\underline{g} = \{g_n\}$ from X to Y are said to be *homotopic* if for every neighborhood V of Y there exists a neighborhood U of V such that

$$f_n/U \simeq g_n/U \text{ in } V \text{ for almost all } n .$$

Evidently, the relation of homotopy for fundamental sequences is reflexive, symmetric and transitive. Hence the collection of all fundamental sequences from X to Y decomposes uniquely into disjoint classes of homotopic fundamental sequences. These classes will be said to be *fundamental classes from* X *to* Y. Let us denote shortly the fundamental class with a representative $\underline{f} = \{f_n\}$ by $[\underline{f}]$. We will write $[\underline{f}]: X \to Y$.

§2. FUNDAMENTAL CLASSES GENERATED BY MAPS

Let us observe that the notion of the fundamental class from X to Y may be considered as a generalization of the notion of the homotopy class of maps of X into Y. In fact, if $\phi: X \to Y$ is a map, then there exists a map $f: H \to H$ such that $f(x) = \phi(x)$ for every point $x \in X$. Setting $f_n = f$ for every $n = 1, 2, \dots$, we get a sequence $\{f_n\}$ of maps which is a fundamental sequence from X to Y.

One proves easily that if $\phi': X \to Y$ is another map homotopic to ϕ in every neighborhood V of Y and if f' is any map of the space H into itself satisfying the condition $f'(x) = \phi'(x)$ for every $x \in X$, then setting $f'_n = f'$ for every $n = 1, 2, \dots$ one gets a fundamental sequence $\{f'_n\}$ from X to Y homotopic to the fundamental sequence $\{f_n\}$.

Thus, if we denote, for every map $\phi: X \to Y$, by $[\phi]_w$ the class of all maps of X into Y homotopic to ϕ in every neighborhood of Y (let us call such a class $[\phi]_w$ a *weak homotopy class* of maps of X into Y), then to every such class $[\phi]_w$ corresponds uniquely a fundamental class $[\underline{f}]: X \to Y$. Let us say that this *fundamental class is generated by the map* ϕ. In particular, the fundamental class from X to X generated by the identity map $i: X \to X$ is said to be the *fundamental identity class for* X.

Manifestly, in general, there exist fundamental classes which are not generated by any map. However, in the case when Y is an ANR-space, one proves easily that the notion of the weak homotopy class is the same as the notion of the homotopy class and every fundamental class $[\underline{f}]: X \to Y$ is generated by a map $\phi: X \to Y$. In this case we have a one-to-one correspondence between homotopy classes of maps of X into Y and the fundamental classes from X to Y. Thus, in this special case the notion of the fundamental class differs only formally from the notion of the homotopy class.

For arbitrary compacta the situation is different and it is to some extent analogous to the situation which we have in the theory of real numbers, as given by Cantor. The weak homotopy classes of maps play the role of the rational numbers, the fundamental classes–the role of real numbers.

§3. COMPOSITION OF FUNDAMENTAL CLASSES

Let X, Y, Z be three compacta lying in the Hilbert space H and let $\{f_n\}$ be a fundamental sequence from X to Y, and $\{g_n\}$, a fundamental sequence from Y to Z. One sees at once that the maps $g_n f_n: H \to H$ constitute a fundamental sequence from X to Z. Moreover, if $\{f_n'\}$ is another fundamental sequence from X to Y homotopic to the sequence $\{f_n\}$, and if $\{g_n'\}$ is a fundamental sequence from Y to Z homotopic to the sequence $\{g_n\}$, then the fundamental sequences $\{g_n f_n\}$ and $\{g_n' f_n'\}$ are homotopic. It follows that the fundamental class from X to Z with the representative $\{g_n f_n\}$ depends only on the fundamental classes $[\underline{f}]$ with the representative $\{f_n\}$ and $[\underline{g}]$ with the representative $\{g_n\}$. We will denote this fundamental class by $[\underline{g}][\underline{f}]$ and we will call it the *composition* of the fundamental classes $[\underline{f}]$ and $[\underline{g}]$. Evidently, if the fundamental class $[\underline{f}]$ is generated by a map $\phi: X \to Y$ and the fundamental class $[\underline{g}]$ is generated by a map $\psi: Y \to Z$, then the composition $[\underline{g}][\underline{f}]: X \to Z$ is generated by the map $\psi\phi: X \to Z$.

One sees at once that, for any three fundamental classes $[\underline{f}]$, $[\underline{g}]$, $[\underline{h}]$, the composition $[\underline{h}]([\underline{g}][\underline{f}])$ is defined if and only if $([\underline{h}][\underline{g}])[\underline{f}]$ is defined.

In this case the associative law holds: $[\underline{h}]([\underline{g}][\underline{f}]) = ([\underline{h}][\underline{g}])[\underline{f}]$. One denotes this triple composition by $[\underline{h}][\underline{g}][\underline{f}]$. It is defined if and only if both compositions $[\underline{g}][\underline{f}]$ and $[\underline{h}][\underline{g}]$ are defined. Moreover, one sees easily that if $[\underline{i}_Y]: Y \to Y$ is the fundamental identity class for Y, then for each fundamental class $[\underline{f}]: X \to Y$ the composition $[\underline{i}_Y][\underline{f}]: X \to Y$ coincides with $[\underline{f}]$, and for every fundamental class $[\underline{g}]: Y \to X$ the composition $[\underline{g}][\underline{i}_Y]: Y \to X$ coincides with $[\underline{g}]$. It follows that we obtain a category if we consider the compacta lying in H as objects and the fundamental classes as mappings. Let us call this category the *fundamental category*.

§4. FUNDAMENTAL DOMINATION AND FUNDAMENTAL EQUIVALENCY

A fundamental class $[\underline{g}]: Y \to X$ is said to be a *right inverse* of the fundamental class $[\underline{f}]: X \to Y$ if the composition $[\underline{f}][\underline{g}]: X \to Y$ is the fundamental identity class for Y. The fundamental classes $[\underline{f}]: X \to Y$ for which there exists a right inverse will be said to be *rightly inversible*. If there exists a rightly inversible fundamental class $[\underline{f}]: X \to Y$, then we say that the compactum X *fundamentally dominates* the compactum Y.

Let us notice that the relation of the fundamental domination does not depend on the position of compacta in the Hilbert space H. In fact, if X', Y', are two compacta in H homeomorphic to X and to Y respectively, then one proves easily that if X fundamentally dominates Y, then X' fundamentally dominates Y'.

Moreover, let us observe that the relation of the fundamental domination is a generalization of the relation of the homotopy domination in the sense of J. H. C. Whitehead ([5], p. 1133). In fact, if X homotopically dominates Y, then there exists a map $f: X \to Y$ and a map $g: Y \to X$ such that $fg: Y \to Y$ is homotopic to the identity. It follows at once, that the fundamental class $[\underline{g}]: Y \to X$ generated by the map g is a right inverse of the fundamental class $[\underline{f}]: X \to Y$ generated by the map f. Hence X fundamentally dominates Y.

One proves easily that if $[\underline{f}]: X \to Y$ and $[\underline{g}]: Y \to Z$ are rightly inversible fundamental classes, then the composition $[\underline{g}][\underline{f}]: X \to Z$ is also a rightly inversible fundamental class. Consequently the relation of the fundamental domination is transitive. Manifestly it is also reflexive, but in general it is not symmetric. It allows to introduce a classification of compacta based on their global topological properties.

A fundamental class $[\underline{f}]: X \to Y$ will be said to be *inversible* if there exists a fundamental class $[\underline{g}]: Y \to X$ such that both compositions $[\underline{f}][\underline{g}]: Y \to Y$ and $[\underline{g}][\underline{f}]: X \to X$ are fundamental identity classes. Then $[\underline{g}]$ will be called the inverse of $[\underline{f}]$. Let us observe that, for every fundamental class $[\underline{f}]: X \to Y$, there exists at most one inverse fundamental class $[\underline{g}]: Y \to X$. In fact, if $[\underline{g}']: Y \to X$ is also an inverse of $[\underline{f}]$, then $[\underline{g}'] = [\underline{g}']([\underline{f}][\underline{g}]) = ([\underline{g}'][\underline{f}])[\underline{g}] = [\underline{g}]$.

Now we can generalize the classical notion of the homotopy type due to W. Hurewicz ([3], p. 525) as follows: Two compacta $X, Y \subset H$ are fundamentally equivalent if there exists an inversible fundamental class $[\underline{f}]: X \to Y$. One sees at once that this relation is reflexive, symmetric and transitive. Hence the collection of all compacta lying in the Hilbert space H decomposes uniquely into disjoint classes of fundamentally equivalent compacta. These classes will be said to be *fundamental types*. It is clear that two compacta of the same homotopy type belong to the same fundamental type. The converse is not true for arbitrary compacta, but it is true for ANR-spaces. Two ANR-spaces of the same fundamental type have necessarily the same homotopy type.

The following example illustrates to some extent the sense of these notions. Let X and Y be two continua lying in the Euclidean plane E^2, which we consider as embedded in the Hilbert space H. If $E^2 - X$ and $E^2 - Y$ have the same number of components, then X and Y are of the same fundamental type. It follows that the collection of all fundamental types of plane continua is only countable. However one shows easily that the cardinality of the collection of all homotopy types of plane compacta is $2^{\aleph_0}$.

§4a. EXTENSION OF FUNDAMENTAL SEQUENCES FUNDAMENTAL RETRACTIONS

Let $\underline{f} = \{f_n\}$ be a fundamental sequence from X to Y and $\underline{f}' = \{f'_n\}$ a fundamental sequence from X′ to Y. If $X \subset X'$ and

$$f_n(x) = f'_n(x) \text{ for every point } x \in X$$

then the fundamental sequence $\underline{f}$ is said to be a *restriction* (to X) of the fundamental sequence $\underline{f}'$, and the fundamental sequence $\underline{f}'$ is said to be an *extension* (onto X′) of the fundamental sequence $\underline{f}$.

One can prove that the existence of an extension of a fundamental sequence $\underline{f}$ from X to Y onto a compactum $X' \supset X$ depends only on the fundamental class of $\underline{f}$.

The notion of the extension for fundamental sequences allows to extend the notion of the retraction. A fundamental sequence $\underline{r} = \{r_n\}$ from X to $Y \subset X$ is said to be a *fundamental retraction of* X *to* Y, if $r_n(y) = y$ for every point $y \in Y$. Then Y is said to be a *fundamental retract of* X. Manifestly, if $\underline{r}$ is a fundamental retraction of X to Y and $\underline{f}$ is a fundamental sequence from Y to Z, then the composition $\underline{f}\,\underline{r}$ is an extension of the fundamental sequence $\underline{f}$ onto X.

Now we can extend also the notions of the absolute retract and of the absolute neighborhood retract. A compactum Y is said to be a *fundamental absolute retract* (FAR) provided it is a fundamental retract of every compactum $X \supset Y$, and Y is said to be a *fundamental absolute neighborhood retract* (FANR) provided for every compactum $X \supset Y$ there exists a compact neighborhood Z of Y in X such that Y is a fundamental retract of Z. Among plane sets, FAR-sets are the same as continua which do not decompose the plane and the FANR-sets are the same as compacta with finite number of components which decompose the plane into a finite number of regions.

Many theorems on AR-sets and ANR-sets can be extended onto FAR-sets and FANR-sets.

§5. FUNDAMENTAL CLASSES AND HOMOLOGY GROUPS

The classical definition of the homology groups of a compactum X, as given by L. Vietoris ([4]), is based on the notions of the ε-simplex and of the true cycle. Let us recall shortly these definitions. By an *n-dimensional ε-simplex in* X (where ε is a positive number) one understands a system $a_0, a_1, \ldots, a_n$ of points of X with diameter less than ε. The linear forms $\kappa = \alpha_1 \sigma_1 + \alpha_2 \sigma_2 + \cdots + \alpha_k \sigma_k$ where σ_i are oriented n-dimensional simplexes in X and α_i are elements of an Abelian group $\mathfrak{A}$, are said to be n-dimensional *chains in* X *over* $\mathfrak{A}$. By a *true* n-*dimensional cycle in* X *over* $\mathfrak{A}$ one understands a sequence $\underline{\gamma} = \{\gamma_k\}$, where γ_k are ε_k-cycles in X over $\mathfrak{A}$ with ε_k converging to zero and if for every k = 1, 2, ... there exists an ε_k-chain in X over $\mathfrak{A}$ having $\gamma_k - \gamma_{k+1}$ as its boundary.

If one defines, in the standard way, the addition of n-dimensional true cycles in X over $\mathfrak{A}$ and the relation of the homology for them, then one gets the group $Z_n(X, \mathfrak{A})$ of n-dimensional true cycles in X over $\mathfrak{A}$ and its subgroup $B_n(X, \mathfrak{A})$ consisting of all true cycles homologous to zero in X. The factor group $Z_n(X, \mathfrak{A})/B_n(X, \mathfrak{A})$ is the n-dimensional homology group $H_n(X, \mathfrak{A})$ of the space X over $\mathfrak{A}$.

If X is a subset of the Hilbert space H, then let us modify the basic definition of the ε-simplex $(a_0, a_1, \ldots, a_n)$ in X so that, instead of the hypothesis $a_i \in X$, let us assume that $a_i \in H$ and that $\rho(a_i, X) < \varepsilon$. One easily sees that this modification is rather unessential and it leads to the same homology groups of X as the classical definition of Vietoris.

Now, if we have two compacta X and Y lying in the Hilbert space H and if $\underline{f} = \{f_k\}$ is a fundamental sequence from X to Y, then it is easy to prove that, for every n-dimensional true cycle (in the modified sense) $\gamma = \{\gamma_i\}$, there exists an increasing sequence $\{i_k\}$ of indices such that the sequence $\{f_k(\gamma_{i_k})\}$ is an n-dimensional true cycle in Y over $\mathfrak{A}$. Moreover, one proves that the homology class of this true cycle doe not depend on the choice of the sequence $\{i_k\}$ and it does not change if one replaces the true cycle $\underline{\gamma}$ by another true cycle $\underline{\gamma}'$ homologous to $\underline{\gamma}$ in X, or if one replaces the fundamental sequence $\{f_k\}$ by another representative of its fundamental class. It follows that if $[\underline{\gamma}]$ is an element of the homology group $H_n(X, \mathfrak{A})$ with the representative $\underline{\gamma} = \{\gamma_i\}$, and if the fundamental sequence $\{f_k\}$ is a representative of the fundamental class $[\underline{f}]$: $X \to Y$, then assigning to $[\underline{\gamma}]$ the element of the homology group $H_n(Y, \mathfrak{A})$ with the representative $\{f_k(\gamma_{i_k})\}$ one obtains an uniquely defined function mapping the group $H_n(X, \mathfrak{A})$ into the group $H_n(Y, \mathfrak{A})$. It is evident that this function is additive, hence it is a homomorphism, which we call the *homomorphism induced by the fundamental class* $[\underline{f}]$. Let us denote it by $[\underline{f}]_*$. Hence

$$[\underline{f}]_*: H_n(X, \mathfrak{A}) \to H_n(Y, \mathfrak{A});$$

One sees easily that

1) if $[\underline{f}]$: $X \to Y$ and $[\underline{g}]$: $Y \to Z$ are fundamental classes, then their composition $[\underline{g}][\underline{f}]$: $X \to Z$ induces the homomorphism

$$([\underline{g}][\underline{f}])_* = [\underline{g}]_*[\underline{f}]_* \quad ;$$

2) if ϕ: $X \to Y$ is a map, then the fundamental class $[\underline{f}]$: $X \to Y$ generated by ϕ induces the same homomorphism $[\underline{f}]_*$: $H_n(X, \mathfrak{A}) \to H_n(Y, \mathfrak{A})$ as the map ϕ. In particular, the fundamental identity classes induce the identity homomorphisms.

It follows from 1) and 2) that

3) if X fundamentally dominates Y, then the group $H_n(Y, \mathfrak{A})$ is isomorphic to a direct factor of the group $H_n(X, \mathfrak{A})$, i.e., there exists a group $\mathfrak{B}$ (depending on n and $\mathfrak{A}$) such that $H_n(X, \mathfrak{A})$ is isomorphic to $H_n(Y, \mathfrak{A}) \times \mathfrak{B}$.

4) if X and Y are fundamentally equivalent, then the group $H_n(X, \mathfrak{A})$

is isomorphic to the group $H_n(Y, \mathfrak{A})$.

Thus we infer that, if one assigns to every fundamental class $[\underline{f}]$ from X to Y the induced homomorphism

$$[\underline{f}]_*: H_n(X, \mathfrak{A}) \to H_n(Y, \mathfrak{A}),$$

then one gets a *covariant functor* H_n from the fundamental category to the category of Abelian groups.

§6. RELATIVIZATION

As it was already remarked, the notion of the fundamental class may be considered as a generalization of the notion of the weak homotopy class. For the sake of brevity, I limited myself to the case of fundamental classes from one compactum X to another compactum Y. But a systematic development of the theory requires the extension of the basic notions onto the case where instead of two compacta lying in the Hilbert space H, one considers two pairs of compacta (X, X_0) and (Y, Y_0) lying in H.

A pair (U, U_0) of subsets of H will be said to be a *neighborhood of the pair* (X, X_0) provided U is a neighborhood of X and U_0 is a neighborhood of X_0. By a fundamental sequence from (X, X_0) to (Y, Y_0) one understands a sequence of maps $f_n: H \to H$ such that for every neighborhood (V, V_0) of the pair (Y, Y_0) there exists a neighborhood (U, U_0) of the pair (X, X_0) such that for almost all n the formula

$$\phi_n(x) = f_n(x) \text{ for every point } x \in U$$

defines a map $\phi_n: (U, U_0) \to (V, V_0)$ and that ϕ_n is homotopic to ϕ_{n+1} in (V, V_0).

Starting from this definition, one extends easily the notions of the fundamental class, of the fundamental equivalence and so on, from the case of compacta to the case of pairs of compacta. One shows in particular, that a fundamental class $[\underline{f}]: (X, X_0) \to (Y, Y_0)$ induces a homomorphism $[\underline{f}]_*: H_n(X, X_0, \mathfrak{A}) \to H_n(Y, Y_0, \mathfrak{A})$ of each homology group of the pair

(X, X_0) into the corresponding group of the pair (Y, Y_0). The homomorphism induced by the composition of two fundamental classes is the composition of the homomorphisms induced by these classes.

§7. HOMOTOPY GROUPS

Let us add some remarks concerning the relations between the homotopy groups induced by the fundamental classes. It is evident, that the classical notion of the homotopy groups is not suitable for this aim. In fact, there exists a plane continuum X decomposing the plane into exactly two regions and such that $\pi_1(X)$ is trivial. As we have already remarked, X is fundamentally equivalent to a circle Y, for which the group $\pi_1(Y)$ is trivial. Thus we have two compacta of the same fundamental type with different first homotopy groups.

However, there exists another notion of the homotopy groups, which is related to the classical one so as the notion of the homology groups in the sense of Vietoris or of Čech is related to the notion of the singular homology groups. An exposition of the theory of so modified homotopy groups, based on the ideas of Čech, was given in 1944 by D. E. Christie [2]. We proceed here by another way, not using the notion of the net homotopies considered by Christie, but using the operation of the homotopy join (see [1], p 46) applied, instead of to maps (as in the classical theory of homotopy groups) to the so-called approximative maps. The notion of the approximative maps, being in fact only a slight modification of the notion of "mappings towards a space" considered by Christie, may be defined as follows:

Let X, Y be two compacta lying in the Hilbert space H. By an *approximative map of* X *towards* Y we understand a sequence $\{\phi_n\}$ of maps $\phi_n\colon X \to H$ such that for every neighborhood V of Y in H the relation $\phi_n \simeq \phi_{n+1}$ in V holds for almost all n. Two approximative maps $\{\phi_n\}$, $\{\psi_n\}\colon X \to Y$ are said to be *homotopic* if for every neighborhood V of Y the relation $\phi_n \simeq \psi_n$ in V holds for almost all n. The collection of all approximative maps of X towards Y decomposes uniquely into disjoint classes of homotopic approximative maps. These classes will be said to be *approximative classes of* X *towards* Y.

It is easy to extend the notion of approximative classes onto the case of pairs of compacta. In particular, if x_0 is a point of the n-dimensional sphere S^n and Y_0 is a point of a compactum $Y \subset H$, then it is easy to define the homotopy join of every two approximative classes of the pair (S^n, x_0) towards the pair (Y, y_0). The approximative classes of (S^n, x_0) towards (Y, y_0) with the operation of the homotopy join constitute a group $\bar{\pi}_n(Y, y_0)$ which may be considered as the required modification of the classical notion of the n-th homotopy group $\pi_n(Y, y_0)$. One shows, that if Y is an ANR-space, the group $\bar{\pi}_n(Y, y_0)$ is isomorphic to the n-th homotopy group $\pi_n(Y, y_0)$.

One can prove that every fundamental class $[\underline{f}]: (X, x_0) \to (Y, y_0)$ induces a homomorphism $[\underline{f}]_*: \bar{\pi}_n(X, x_0) \to \bar{\pi}_n(Y, y_0)$ and that the composition of two fundamental classes induces the homomorphism being the composition of the homomorphisms induced by these fundamental classes. Since the fundamental identity classes induce the identity homomorphisms, we infer that the corresponding homotopy groups of two fundamentally equiv-equivalent compacta are isomorphic.

Finally, let us remark, that if one assigns to each fundamental class $[\underline{f}]$ from (X, x_0) to (Y, y_0) the induced homomorphism

$$[\underline{f}]_*: \bar{\pi}_n(X, x_0) \to \bar{\pi}_n(Y, y_0),$$

then one gets a covariant functor Π_n from the fundamental category to the category of groups (Abelian, if $n > 1$).

If $x_0 = y_0 \in Y \subset X$ and $\underline{f}$ is a fundamental retraction of X to Y, then one proves easily that the homomorphisms of the homology and homotopy groups induced by $\underline{f}$ are so-called r-*homomorphisms* (see [1], p. 32). It follows in particular that all homology and homotopy groups of a fundamental absolute retract are trivial, and the homology and homotopy groups of a fundamental absolute neighborhood retract are r-images of the corresponding groups of a polyhedron.

REFERENCES

[1] K. Borsuk, Theory of Retracts, *Monografie Matematyezne 44*, Warszawa 1967.

[2] D. E. Christie, Net homotopy for compacta, *Trans. Amer. Math Soc.*, *56* (1944), pp. 275-308.

[3] W. Hurewicz, Beiträge zur Topologie der Deformationen III, *Proc. Ak. Amsterdam, 39* (1936), pp. 117-125.

[4] L. Vietoris, Über den höheren Zusammenhang kompakter Räume und eine Klasse von zusammenhangstreuen Abbildungen, *Math. Ann.*, *97* (1927), pp. 454-472.

[5] J. H. C. Whitehead, On the homotopy type of ANR's, *Bull. Amer. Math. Soc.*, *54* (1948), pp. 1133-1145.

Warsaw

SOME THIN SETS IN FRÉCHET SPACES

H. H. CORSON

Let s be the countable product of real lines, and let ℓ_2 denote all square summable sequences. It has been known for some time that the complement of a point in ℓ_2 is homeomorphic to ℓ_2, with a similar result holding for s. This follows from results of O. H. Keller, V. Klee, and R. D. Anderson. In order to prove that the complements of other subsets of these spaces are homeomorphic to the original space, one approach has been to straighten out these subsets with respect to a topologically complemented subspace and then remove one point at a time. We will show how this straightening can be accomplished in a linear way which depends only on the classical theorems of functional analysis, in contrast to the other methods which are *ad hoc* topological techniques. This follows a suggestion of C. Bessaga at the L. S. U. conference that it might be interesting to prove results of this type by the methods of functional analysis. (See [1] for the necessary theorems used below.)

Our results hold in the general context of Fréchet spaces, that is, complete metric locally convex vector spaces.

THEOREM. *Let* X *be a Fréchet space with infinite dimensional closed subspaces* $E_1, E_2, \dots$. *Let* K *be a* σ*-compact subspace of* X. *Then there are linearly independent points* $x_i \in E_i$ *such that the closed linear span of* $\{x_i\}$ *has only* 0 *in common with* K.

Proof: Since X is locally convex and $K \setminus \{0\}$ has the Lindelöf property, we can find a countable collection $\{K_i\}$ of compact convex subsets of $X \setminus \{0\}$ which cover $K \setminus \{0\}$. Hence we may assume that $K_i \subset K$, all i.

By induction, we will show that one may find closed subspaces X_i of finite co-dimension in X and points $x_i \in X_i \cap E_i$ with these properties:

(a) $X_{i+1} \subset X_i \setminus \{x_i\}$,

(b) $Y_i \cap K = \{0\}$ where Y_i is the linear span of $\{x_1, x_2, \ldots, x_i\}$, and

(c) $K_{i+1} + Y_i$ does not meet X_{i+1}, where $Y_0 = \{0\}$.

In fact, X_1 can be chosen as the null space of a continuous linear functional separating 0 from K_1, by the separation theorem. Then $X_1 \cap E_1$ is closed and infinite dimensional, hence not locally compact; so a category argument shows that there is a point x_1 in $(X_1 \cap E_1) \setminus K$.

Now suppose that $x_1, \ldots, x_m$ and $X_1, \ldots, X_m$ are chosen. It is easy to see that $Y_m + K_{m+1}$ is closed and convex, and by (b) $0 \notin Y_m + K_{m+1}$. Again by the separation theorem one may pick X_{m+1} of finite co-dimension such that X_{m+1} does not meet $Y_m + K_{m+1}$, and we may obviously suppose that $X_{m+1} \subset X_m \setminus \{x_m\}$. As above, $(X_{m+1} \cap E_{m+1}) \setminus (K + Y_m)$ contains some point x_{m+1}. Clearly (a) and (c) are valid. For (b) suppose that $ax_{m+1} + \Sigma_i^m a_i x_i \in K \setminus \{0\}$. Then $a \neq 0$ for otherwise $Y_m \cap K \neq \{0\}$. If $a \neq 0$, then x_{m+1} would be in $K + Y_m$ which is not the case.

To complete the proof, suppose that $y \in K \setminus \{0\}$ and

$$y = \lim_{m \to \infty} \sum_i a_i^m x_i \ ,$$

where each of the sums is finite. By the Hahn-Banach theorem there is a continuous linear map of X onto Y_i such that X_{i+1} goes to 0, since (a) implies that $X_{i+1} \cap Y_i = \{0\}$. Since (a) also implies that $\{x_1, \ldots, x_i\}$ is a set of linearly independent points in X, it follows that $a_i = \lim_{m \to \infty} a_i^m$ exists for each i.

Suppose that $y \in K_{i_0}$. Then $y = a_1 x_1 + \cdots + a_{i-1} x_{i-1} + z$ where

$$z = \lim_{m \to \infty} \sum_{i \geq i_0} a_i^m x_i \in X_{i_0} \ .$$

It follows that $z \in K_{i_0} + Y_{i_0 - 1}$, contradicting (c) and completing the proof.

COROLLARY 1. *Given a* σ*-compact subset* K *of* ℓ_2 there is an infi*nite dimensional closed subspace* H *of* ℓ_2 *and a complementary subspace* H' *of* ℓ_2, *each isomorphic to* ℓ_2, *such that each translate of* H *meets* K *in at most one point.*

Proof: We may suppose that K is a linear subspace. Pick any infinite dimensional closed subspace E of ℓ_2 such that E is of infinite co-dimension. Let $E_i = E$ for all i. Let H be the closed linear span of $\{x_i\}$ of the Theorem, and let H' be its orthogonal complement. Then it is well-known that H and H' are isomorphic to ℓ_2. Clearly H has the desired property.

COROLLARY 2. *Replace* ℓ_2 *by* s *in Corollary* 1.

Proof: Partition the integers into infinite subsets $N_1, N_2, \ldots$. Let E_i be the subspace of s composed of all sequences whose support is contained in N_i. Let H be the closed linear span of the $\{x_i\}$ of the theorem. Let E_i' be a closed proper subspace of E_i such that E_i' is isomorphic to s and $E_i' \cup \{x_i\}$ generates E_i. Let H' be the closed linear span of $\cup \{E_i' : i \geq 1\}$.

To prove that H and H' are isomorphic to s it is only necessary to observe that E_i is isomorphic to $E_i' \times \{ax_i : a \text{ real}\}$ and s is isomorphic to $\Pi_{i=1}^{\infty} E_i$. Clearly H has the desired property.

COMMENT. For the topological applications of the straightening theorems it is sometimes enough to know that X is homeomorphic to $X_1 \times X_2$ where X_i are infinite dimensional Fréchet spaces and the image of K in $X_1 \times X_2$ meets each translate of X_2 in at most one point. This can always be accomplished by applying the above Theorem together with the Bartle-Graves theorem [2]. It does not seem likely, however, that one can assert that there is an isomorphism of X into some $X_1 \times X_2$ and *also* specify the linear properties of X_1 and X_2 in terms of those of X, as in Corollaries 1 and 2.

Question. Can one always find an isomorphism as in the Comment above for some X_1 and X_2 ? The answer is affirmative if every infinite dimensional closed subspace contains an infinite dimensional closed subspace which is complemented in X.

Note. If $E_i = X$ for all i and X is *not* a reflexive Banach space, then the conclusion of the theorem holds under the assumption that K is σ-compact in the weak topology.

REFERENCES

[1] M. M. Day, *Normed Linear Spaces,* Berlin, Springer-Verlag, 1958.

[2] E. Michael, Continuous selections, *J. Ann. of Math. (2), 63* (1956), 361-382.

UNIVERSITY OF WASHINGTON, SEATTLE

A REMARK ON BANACH ANALYTIC SPACES

A. Douady
(Faculté des Sciences, Nice)

L' Universite mene a tout,
à condition d'en sortir.
(French proverb.)

In [1], Banach analytic spaces were used as a tool for a moduli problem in complex analytic geometry. However, the only interesting aspect of these spaces is their use as intermediates in constructing an analytic space which eventually turns out to be (locally) finite dimensional. I believe that there is no point in studying Banach analytic spaces for their own sake. For instance, the fact that a topological space is homeomorphic to a Banach analytic space gives no information about its topological type. It is likely that any complete metric space is homeomorphic to a Banach analytic space.

Let us prove:

PROPOSITION. *Any compact metric space is homeomorphic to a Banach analytic space.*

Proof. Let X be a compact metric space. Denote by d its distance and by A the Banach algebra of Lipschitz functions from X to $\mathbb{C}$ provided with the norm

$$\|f\| = \sup_{x \in X} |f(x)| + \sup_{\substack{x \in X \\ y \in X \\ x \neq y}} \frac{|f(x) - f(y)|}{d(x, y)} .$$

Consider the Banach space A' dual to A with the norm topology, and the map $\delta : X \to A'$ defined by

$$<f, \delta(x)> = f(x), \quad f \in A .$$

We have the inequalities

$$\frac{d(x,y)}{1+d(x,y)} \leq \|\delta(x) - \delta(y)\| < d(x,y), \quad x, y \in X ;$$

the second one is obvious and the first one is obtained by considering the function on X

$$z \mapsto \inf(d(x,z),\ d(x,y)) .$$

These inequalities show that δ is a homeomorphism of X onto its image in A'.

It is well-known that $\delta(X)$ is the spectrum of A, i.e., the set of algebra homomorphism from A to $\mathbb{C}$. In other words, let B denote the Banach space of bilinear maps from $A \times A$ to $\mathbb{C}$, and ϕ the map from A' to $B \oplus \mathbb{C}$ defined by $\phi(\xi) = (\theta, \phi(1) - 1)$, where

$$\theta(f,g) = \xi(fg) - \xi(f)\xi(g), \quad f, g \in A .$$

Then ϕ is analytic and $\delta(X) = \phi^{-1}(0)$.
Therefore $\delta(X)$ is a Banach analytic subspace of A'.

q.e.d.

STANFORD UNIVERSITY
FEBRUARY 1968

REFERENCE

[1] A. Douady, *Le problème des modules pour les sous-espaces analytiques compacts d'un espace analytique donné*, Ann. Inst. Fourier (Grenoble) 16 (1966), 1-95.

FIBRING SPACES OF MAPS

James Eells, Jr.

1. *Introduction*

This article summarizes briefly the role played by fibre spaces in the structure theory of various spaces and manifolds of maps. We consider both topological and differential aspects. In particular, we are concerned with conditions sufficient to insure (1) that certain maps satisfying the covering homotopy property are locally trivial; (2) that certain differentiable maps which foliate a differentiable manifold actually define a locally trivial fibration; (3) that certain locally trivial fibrations are (or are not) in fact fibre bundles associated with continuous topological (or Lie) group actions.

Special attention is drawn to three important classes of applications: (1) The evaluation map fibrations of Serre and Borsuk, in their topological framework. (2) The fibrations arising in the embedding and immersion theory of Cerf and Hirsch-Smale. (3) Applications (due to Earle-Eells) to Teichmüller theory, which produce new methods and results in the global theory of deformations of complex structures (of complex dimension 1).

Throughout, our notation and terminology will conform to that of [11] – to which we refer also for background theory.

2. *Fibrations and fibre bundles*

Let us begin with a review of the topological theory of fibre bundles; the basic references here are Cartan [2], Dold [6], Ehresmann [12], Steenrod [27]. Our viewpoint is more that of [2], [12] than of [27]. We state three fundamental theorems on which the topological theory of fibre bundles is based.

(A) Let E be a topological space and G a topological group *acting continuously on the right of* E; i.e., we have a continuous map $E \times G \to E$ written $(x,g) \to x \cdot g$, such that $x \cdot (g_1 g_2) = (x \cdot g_1) \cdot g_2$ and $x \cdot 1 = x$ for all $x \in E$, $g_1, g_2 \in G$, where 1 denotes the neutral element of G. We let $B = E/G$ denote the orbit space and $p: E \to B$ the surjective map assigning to each $x \in E$ its orbit $p(x) \in B$ under the G-action; we give B the quotient topology (i.e., the largest topology for which p is continuous). Suppose that G acts *freely* (i.e., we have $x \cdot g = x$ only when $g = 1$); then setting $\Delta = \{(x,x') \in E \times E$: there is a (unique) $g \in G$ for which $x' = x \cdot g\}$, we have a map $\theta: \Delta \to G$ defined by $\theta(x, x \cdot g) = g$, and we say that the G-action is *proper* if θ is continuous. Finally, if every point $b \in B$ has a neighborhood V_b on which there is a continuous section ζ_b (i.e., a continuous map $\zeta_b : V_b \to E$ such that $p \cdot \zeta_b$ is the identity map on V_b), then we say that the G-action is *locally trivial.* A *principal right* G*-action* is one which is continuous, free, proper, and locally trivial. We then say that $p: E \to B$ is a *principal fibre bundle with structural group* G; or a G-*bundle*, for short.

Next, suppose that F is a topological space on which G acts continuously on the left. Then G acts principally on $E \times F$ by $(x,f) \cdot g = (x \cdot g, g^{-1} f)$; we let $W = E \times_g F$ denote its orbit space. There is a natural surjective continuous map $q: W \to B$, and we call it a *fibre bundle with structural group* G *and fibre model* F, *associated to the* G-*bundle* $p: E \to B$; or a (G,F)-*bundle*, for short.

(B) A continuous surjective map $q: W \to B$ of topological spaces is *locally trivial with fibre* F if every $b \in B$ has a neighborhood V over which there is a homeomorphism θ such that the diagram

$$\begin{array}{ccc} q^{-1}(V) & \xrightarrow{\theta} & V \times F \\ & {\scriptstyle q} \searrow \quad \swarrow {\scriptstyle \pi} & \\ & V & \end{array}$$

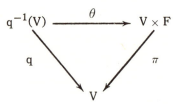

is commutative, where π is the indicated projection on the first factor. We will say that a locally trivial map $q:W \to B$ *defines a fibration of* W. Every (G,F)-bundle defines a fibration.

Henceforth assume that all spaces are Hausdorff.

COVERING HOMOTOPY THEOREM. *Let* $q:W \to B$ *be a fibration of* W *over the paracompact space* B. *Let* P *be a space and* $f_I:P \times I \to B$ *a continuous map.* (*Here* $I = [0,1]$.) *If* $g_0:P \to W$ *is a map such that* $q \cdot g_0 = f_0$, *then there is an extension* $g_I:P \times I \to W$ *such that* $q \cdot g_I = f_I$:

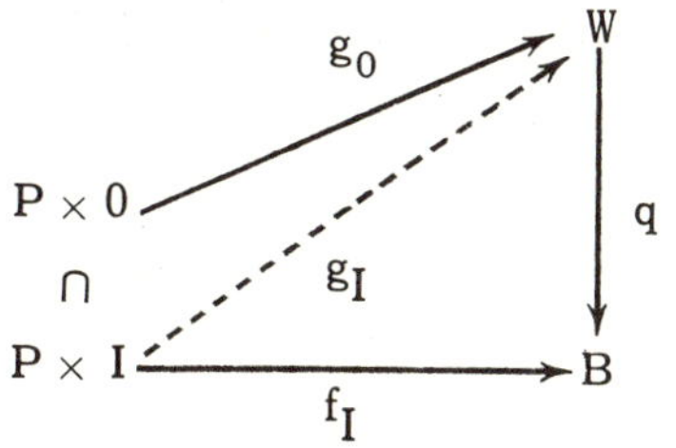

REMARK. An alternative reading of that theorem is valid, requiring P to be paracompact and allowing B to be arbitrary.

A beautiful proof (Hurewicz [18]) of the covering homotopy theorem is based on the existence of a Hurewicz connection: Let $C^0(I,B)$ denote the mapping space of all continuous maps of I into B, with the compact-open topology. Let $q^{-1}C^0(I,B) = \{(x,\omega)\ W \times C^0(I,B): q(x) = \omega(0)\}$:

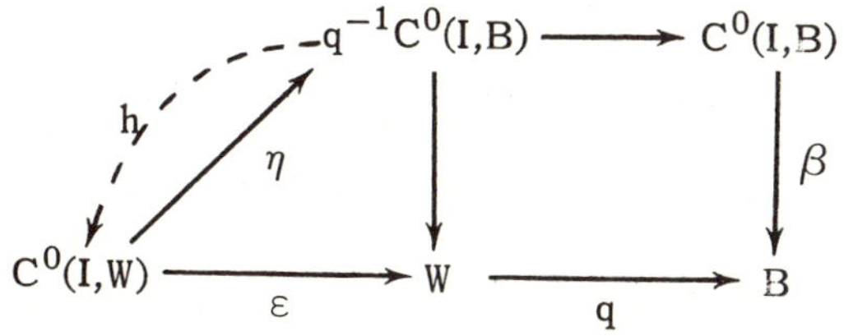

The Serre map $\beta:C^0(I,B) \to B$ is given by $\beta(\omega) = \omega(0)$; similarly for $\varepsilon:C^0(I,W) \to W$. The map η is defined by $\eta(\phi) = (\phi(0), q \cdot \phi)$. A *Hurewicz connection* for $q:W \to B$ is a continuous section h of the map η; it provides a coherent lifting of paths from B to W, with prescribed initial point. It is easy to show that the covering homotopy theorem is valid for all spaces P if and only if $q:W \to B$ admits a Hurewicz connection. Such a map is called a *Hurewicz fibration.*

(C) A map $q:W \to B$ has the *local section extension property* if every point of B has a neighborhood V such that any section of q defined on a closed subset A contained in V has an extension to a neighborhood of A in V.

SECTION EXTENSION THEOREM. *Let* $q:W \to B$ *be a fibration of* W *over the paracompact space* B, *and suppose that* q *has the local section extension property. Assume that the fibre* F *is contractible. If* B_0 *is a closed subspace of* B *and* s_0 *a section of* q *over* B_0, *then there is an extension to a section* s *over all* B.

(D) If G is a topological group, then a G-bundle $p_G:E_G \to B_G$ is G-*universal* if the space E_G is contractible. It is known [23, 6] that every topological group admits a G-universal bundle, which is sufficiently unique for topological classification purposes.

An *isomorphism* of G-bundles over B

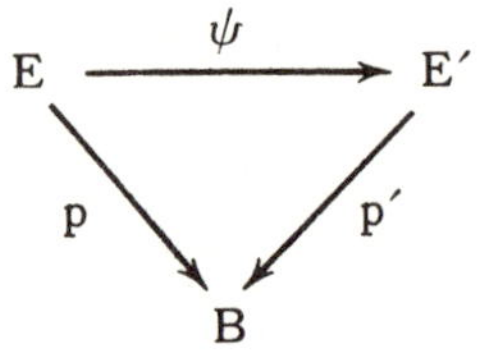

is a homeomorphism ψ which is equivariant (i.e., $\psi(x \cdot g) = \psi(x) \cdot g$ for all $x \in E$, $g \in G$) and which induces the identity map on B.

CLASSIFICATION THEOREM. *Let* G *be a topological group and* $p_G:E_G \to B_G$ *a universal* G-*bundle. Let* B *be a paracompact space. Then the isomorphism classes of* G-*bundles over* B *are in natural bijective correspondence with the homotopy classes of maps* $B \to B_G$.

The proof depends on both the covering homotopy theorem and the section extension theorem.

3. *Foliations and fibrations*

(A) Let X and Y be paracompact C^1-manifolds modeled on Banach spaces, and $f:X \to Y$ a surjective C^1-map. Suppose that for each $x \in X$ its differential $f_*(x):X(x) \to Y(f(x))$ is surjective and that its kernel $\mathrm{Ker}\, f_*(x)$ is a direct summand of $X(x)$. We will then say that f *foliates* X. The components of the sets $f^{-1}(y)$ (for each $y \in Y$) are called the *leaves* of the foliation. The following assertion – which is a consequence of the inverse function theorem – insures that the leaves of a foliation are closed C^1-submanifolds of X:

For each $x \in X$ there are neighborhoods U of x and V of $f(x)$, an open set W in a Banach space isomorphic to $\mathrm{Ker}\, f_*(x)$, and a C^1-diffeomorphism $\psi:U \to V \times W$ such that the following diagram is commutative:

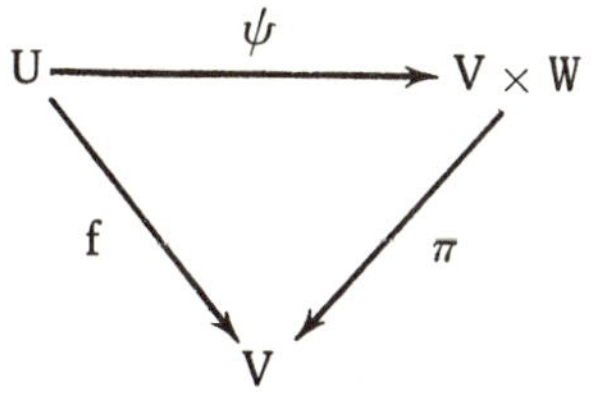

where again π denotes projection on the first factor. An example of Douady [7, Ch. I, §8] shows that the direct summand condition on $\mathrm{Ker}\, f_*(x)$ is essential for the foliation of X by f.

Let $\mathrm{Ker}\, f_* = \cup\{\mathrm{Ker}\, f_*(x) : x \in X\}$. Then $\mathrm{Ker}\, f_*$ is an integrable sub-bundle of the tangent vector bundle $T(X)$ of X, as in [11, §4F], and we have the following exact sequence

$$0 \to \mathrm{Ker}\, f_* \to T(X) \xrightarrow{f_*} f^{-1}T(Y) \to 0 . \tag{1}$$

(B) Let α and β be Finsler structures for X and Y, and suppose that X is complete in the metric induced from α. Let s be a locally Lipschitz splitting of the sequence (1), viewed as a map $s:f^{-1}T(Y) \to T(X)$ which is bounded and linear on the fibres, and such that $f_* \cdot s = 1$. We will say that s *is bounded locally over* Y if for each $y_0 \in Y$ there is a number $\eta_0 > 0$ and a neighborhood V_0 of y_0 such that $\|s(x)\|_x \le \eta_0$ for all $x \in f^{-1}(V_0)$, where

$$\|s(x)\|_x = \sup\{\alpha_x(s(x)v)/\beta_{f(x)}(v) : v \neq 0 \ \text{ in } \ Y(f(x))\}\ .$$

The following result is the main object of [9]:

THEOREM. *Let* (X,α), (Y,β) *be Finsler* C^1*-manifolds modeled on Banach spaces, and suppose that* (X,α) *is complete. Let* $f:X \to Y$ *be a surjective* C^1*-map which foliates* X. *If there is a locally Lipschitz splitting of the sequence* (1) *which is bounded locally over* Y, *then* $f:Y \to Y$ *is a locally* C^0*-trivial fibration.*

The idea of the proof is to use the splitting and the fundamental theorem of ordinary differential equations in Banach manifolds to construct a sort of Hurewicz connection h:

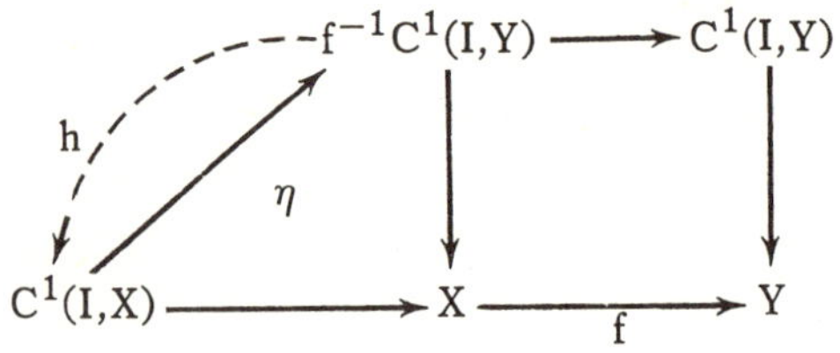

EXAMPLE (Ehresmann [12]). If $f:X \to Y$ foliates X and is proper (i.e., the inverse image of every compact subset of Y is compact in X), then $f:X \to Y$ is a locally C^0-trivial fibration.

EXAMPLE (Hermann [16]). Let X be a separable complete Riemannian C^k-manifold $(1 \leq k \leq \infty)$ and $f:X \to Y$ a surjective C^k-map foliating X. For every $x \in X$ we let $K_x^\perp$ denote the orthogonal complement of $\operatorname{Ker} f_*(x)$. If each $f_*(x)|K_x^\perp \to Y(f(x))$ is an isometry, then $f:X \to Y$ is a locally C^k-trivial fibration. (The theorem admits an easy modification (applicable to this example) to take into account higher differentiability.)

(C) Let (X,α) be a Finsler C^1-manifold and G an abstract group of C^1-diffeomorphisms which are furthermore isometries of the Finsler structure α. Suppose that the G-orbits foliate X. If $Y = X/G$ is the orbit space and $f:X \to Y$ the orbit map, then the quotient topology on Y is given by the metric

$$\tau(y,y') = \inf\{\sigma(x,x') : x \in f^{-1}(y),\ x' \in f^{-1}(y')\}\ ,$$

where σ is the metric on X induced from α. Then Y is itself a Finsler C^1-manifold, and for each $y \in Y$ the quotient Finsler structure β_y is defined as the quotient norm.

The next result is an application of the theorem in §3B.

PROPOSITION. *Let* (X,α) *be a complete Finsler* C^1*-manifold, and* G *a group of* C^1*-diffeomorphisms and isometries of* X. *Suppose that the* G*-orbits foliate* X. *Then the orbit map* $f: X \to Y = X/G$ *is a locally* C^0*-trivial fibration.*

The following illustration is a reformation and specialization of results in [8]:

EXAMPLE. Let X be the open unit disc in $L^\infty(U;\mathbb{C})$, where U is the upper half plane. Let α be the Finsler structure which assigns to each $\mu \in X$ the norm $\nu \to \alpha_\mu(\nu)$ given by

$$\alpha_\mu(\nu) = 2\left\| \frac{\nu}{1-|\mu|^2} \right\|_\infty$$

on $L^\infty(U;\mathbb{C})$. It is easily verified that α is a complete Finsler structure. Let G denote the group of quasi-conformal maps of U onto itself leaving the real axis pointwise fixed. Then G acts as a group of diffeomorphisms of the manifold X and as a group of isometries of the Finsler structure. Furthermore, G foliates X, and each leaf is a complex analytic submanifold. Therefore $f: X \to X/G = Y$ is locally C^0-trivial. (See [8] and [11, §5G] for an explicit formula for f and for its differential.) The quotient space is the *universal Teichmüller space* (for complex dimension one). A similar construction can be made for the Teichmüller space $T(\Gamma)$ of any Fuchsian group Γ.

It is natural to ask whether G can be given a topology so that its action is principal. The answer is no, because then each fibre would be homeomorphic to G, and it is known that G is not a topological group with the induced topology of $L^\infty(U;\mathbb{C})$.

4. *Various examples of fibre bundles*

(A) Let $p:E \to B$ be a locally trivial fibration with locally compact, locally connected fibres F. If we let $\mathcal{C}(F)$ denote the group of all homeomorphisms of F on itself, with the compact-open topology, then $\mathcal{C}(F)$ is a topological group and the evaluation map $\mathcal{C}(F) \times F \to F$ defined by $(g,z) \to g(z)$ is continuous [1, Theorem 4]. In fact, the compact-open topology on $\mathcal{C}(F)$ is the smallest topology for which $\mathcal{C}(F)$ is a topological transformation group of F. Thus *we can view* $p:E \to B$ *as a* $(\mathcal{C}(F), F)$-*bundle.*

Similarly, let $f:X \to Y$ be a locally C^∞-trivial fibration of C^∞-manifolds with finite dimensional fibres with model F. If $\mathcal{D}(F)$ denotes the group of C^∞-diffeomorphisms of F, with the compact-open topology on differentials of all orders, then $\mathcal{D}(F)$ is a topological group and the evaluation map $\mathcal{D}(F) \times F \to F$ is continuous. Thus *we can view* $f:X \to Y$ *as a* $(\mathcal{D}(F), F)$-*bundle.*

There are examples of such bundles having no finite dimensional Lie structural group G; *in particular,* $\mathcal{D}(F)$ *will then have no Lie group as deformation retract.* The following construction has been used for several different purposes [28]:

EXAMPLE (Serre). Let $h:S^7 \to S^4$ be the indicated Hopf fibration; that is an analytic principal $Sp(1) = S^3$-bundle. Let $\phi:S^3 \times S^1 \to S^4$ be a smooth map of degree 1, and $k:M \to S^3 \times S^1$ the induced $Sp(1)$-bundle:

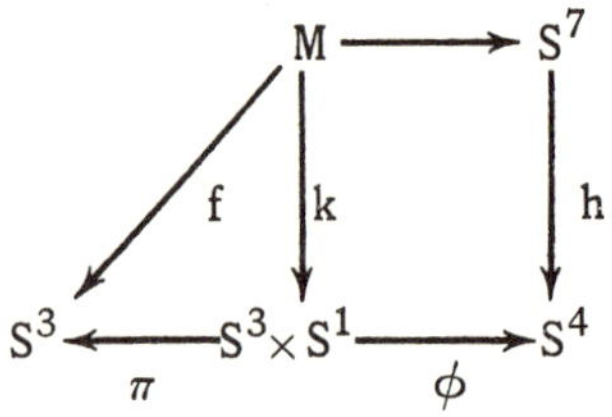

If π denotes projection on the first factor, then the composition $f = \pi \cdot k$ is a locally C^∞-trivial fibration (being a composition of foliations of compact C^∞-manifolds). It is easily seen that its fibre is $S^3 \times S^1$, so that

$f:M \to S^3$ is a $(\mathfrak{D}(S^3 \times S^1), S^3 \times S^1)$-bundle. On the other hand, for any topological group G the isomorphism classes of G-bundles over S^3 are classified [27, §18] by the $\pi_0(G)$-classes of $\pi_2(G)$. For any finite dimensional Lie group G a theorem of E. Cartan asserts that $\pi_2(G) = 0$, so that any (G,F)-bundle over S^3 is trivial. However, $f:M \to S^3$ is not trivial, for (using the Gysin sequence) M is not cohomologically the product of S^3 and $S^3 \times S^1$.

(B) EXAMPLE [10]. Let S be a compact oriented C^∞-manifold of dimension 2, without boundary and of genus $(S) = g \geq 2$. Let $\mathfrak{D}_0(S)$ be the identity component of $\mathfrak{D}(S)$. If M(S) denotes the manifold of all C^∞-complex structures on S, with its C^∞-topology, then $\mathfrak{D}_0(S)$ *operates principally on* M(S), *and the orbit space* $T(S) = M(S)/\mathfrak{D}_0(S)$ *is canonically identified with the Teichmüller space of* S. In fact, if we represent $S = U/\Gamma$ as in the example of §3C, then $T(S) = T(\Gamma)$. Furthermore, since M(S) is contractible, we see that $M(S) \to T(S)$ *is a universal bundle for the topological group* $\mathfrak{D}_0(S)$. It is a theorem of Teichmüller that T(S) is a (3g-3)-$\dim_{\mathbf{C}}$ cell (it is moreover known to be a Stein manifold), so that *every* $\mathfrak{D}_0(S)$*-bundle over a paracompact space is topologically trivial.* Another corollary of this construction (with simple modifications to take care of the exceptional cases $g=0$, $g=1$), together with the fact that $\mathfrak{D}_0(S)$ is an absolute neighborhood retract, is the following: *If* $g=0$, then $\mathfrak{D}(S) = \mathfrak{D}_0(S)$ *has* SO(3) *as deformation retract* (theorem of Smale); *if* $g=1$, *then* $\mathfrak{D}_0(S)$ *has* $S^1 \times S^1$ *as deformation retract; if* $g \geq 2$, *then* $\mathfrak{D}_0(S)$ *is contractible.*

REMARK. The topological groups $\mathcal{C}(S)$ of homeomorphisms of S are more difficult to handle. ($\mathcal{C}(S)$ is a Baire space and locally contractible; but it is not known (to me) whether $\mathcal{C}(S)$ is an absolute neighborhood retract.) It has been shown by Hamstrom [15] that the identity components $\mathcal{C}_0(S)$ of $\mathcal{C}(S)$ have the homotopy groups of the above-mentioned deformation retracts for the $\mathfrak{D}_0(S)$ cases.

(C) If S is a finite dimensional connected manifold and we fix a point $s_0 \in S$ to define the evaluation map $p:\mathcal{C}(S) \to S$ by $p(g) = g(s_0)$, then *we have a principal* $\mathcal{C}_{s_0}(S)$*-bundle, where* $\mathcal{C}_{s_0}(S) = \{g \in \mathcal{C}(S): g(s_0) = s_0\}$; see for instance [21]. Thus S has the representation $\mathcal{C}(S)/\mathcal{C}_{s_0}(S)$ as a homogeneous space.

REMARK. In the case that S is a compact connected C^∞-manifold there is reason to hope that the analogous map $p:\mathfrak{D}(S) \to S$ will possess good differential $\mathfrak{D}_{s_0}(S)$-bundle properties in relation to the exact sequence

$$0 \to C^\infty_{s_0}(T(S)) \to C^\infty(T(S)) \overset{\overline{p}}{\to} S(s_0) \to 0$$

of vector spaces, where $C^\infty_{s_0}(T(S))$ = the space of all C^∞-vector fields on S which vanish at s_0 and $\overline{p}(u) = u(s_0)$; in particular, with respect to the exponential map $C^\infty(T(S)) \to \mathfrak{D}(S)$. As a first step in this direction, Leslie [19] has introduced a differential structure (which we will call of class WC^∞) on $\mathfrak{D}(S)$ relative to which $\mathfrak{D}(S)$ is a WC^∞-manifold and the group operations are WC^∞.

It is known that the exponential map is not locally bijective in the C^∞-topology. However, the WC^∞-structure may still be significant in treating finite dimensional and finite codimensional subspaces of $C^\infty(T(S))$.

(D) EXAMPLE. Let G be a compact group and $p:E \to B$, $p':E' \to B'$ two G-bundles. Assume that B (and hence E) is locally compact, and that B′ is paracompact. Denote by C(B,B′) the space of all continuous maps of B into B′, topologized by the compact-open topology; similarly, let $C_G(E,E')$ denote the space of G-equivariant maps. *There is a natural map* $\eta: C_G(E,E') \to C(B,B')$, *which is a Hurewicz fibration over its image; its fibre model is* $F = \{u: E \to G: u(x\cdot g) = g^{-1}u(x)g$ *for all* $x \in E, g \in G\}$, *with the compact-open topology.* In case $p': E' \to B'$ is a smooth bundle and E is compact we have a differentiable form of that theorem, in terms of manifolds of maps. Also, if B_0 is a closed subspace of B and $p_0: E_0 = p^{-1}(B_0) \to B_0$ is the restriction, then we have the natural map $C_G(E,E') \to C_G(E_0,E')$, which under mild conditions on (B,B_0) is a Hurewicz fibration. See I. M. James, *The space of bundle maps*. Topology 2 (1963), 45-59, where applications are made.

(D) EXAMPLE. Let G be a metrizable topological group and K a closed subgroup. A theorem of Michael [22] asserts that *if* K *is complete and locally convex, there is a local section of the coset map* $G \to G/K$; *therefore we have a homogeneous* K*-bundle*. That is particularly interesting when $G = GL(V)$, the group of linear automorphisms of an infinite dimensional Hilbert space, topologized as an open subset of the Banach algebra $L(V)$ of endomorphisms of V. For then by a theorem of Kuiper G is an absolute retract, and $G \to G/K$ is a universal K-bundle.

Similarly, if V is a separable infinite dimensional Hilbert space and $UL_s(V)$ denotes the group of unitary operators with its strong operator topology, then $UL_s(V)$ is metrizable and contractible [5, §10]. For any locally compact group K we take its Lebesgue space $L^2(V)$ using Haar measure; then K can be imbedded as a closed subgroup of $UL_s(L^2(K))$ – and whenever it has a local section we have a universal K-bundle.

REMARK. It would be interesting to know whether every compact Lie group can operate principally on every C^∞-manifold X modeled on the infinite dimensional separable Hilbert space E. It is true that G *operates principally and smoothly on* E *itself*. For G can be represented faithfully as a closed subgroup of an orthogonal group O_n, which operates principally and analytically on the Stiefel manifold $V_n(E)$ of orthonormal n-frames in E. But by a theorem of Bessaga, coupled with a remark made to me by Husemoller, $V_n(E)$ is C^∞-diffeomorphic to E.

5. *Certain fibrations of mapping spaces*

(A) That certain evaluation maps define a sort of fibration (with sufficient structure to insure the covering homotopy theorem for cells; such a map is called a *Serre fibration*) follows from Borsuk's extension theorem [13, 24]. Such fibrations of path spaces have been used extensively in the celebrated thesis of Serre.

THEOREM [24]. *Let* S *be a compact space and* A *a closed subspace. Then for any absolute neighborhood retract* M *the evaluation map* $\bar{f}\colon C(S,M) \to C(A,M)$ *defined by* $\bar{f}(x) = x|A$ *is a Serre fibration over the image.*

Here $C(S,M)$ denotes the mapping space of all continuous maps of S into M, topologized by the compact-open topology.

A similar theorem with somewhat different hypotheses is the following [17]:

Let S *be a locally compact absolute neighborhood retract and* A *a closed subset which is also an absolute neighborhood retract. Then for any space* M *the evaluation map* $\overline{f}: C(S,M) \to C(A,M)$ *is a Hurewicz fibration over the image.*

Suppose that M is a C^{r+2}-manifold $(r \geq 0)$ modeled on a C^{r+2}-smooth Banach space, and that S is a compact space. Then $C(S,M)$ and $C(A,M)$ are C^r-manifolds; furthermore, then $\overline{f}$ *is a* C^r*-map, and is locally* C^0*-trivial.* If S is metrizable and A has separable frontier, then $\overline{f}$ is a foliation map (see [14, §11]). In that case the sequence (1) becomes

$$0 \to \operatorname{Ker} f_* \to C(S,T(M)) \to \overline{f}^{-1}C(A,T(M)) \to 0 \ ;$$

and we can construct a locally Lipschitz splitting to apply Theorem 3B, giving a differentiable interpretation of that result.

(B) If in the above theorem S and A are compact C^∞-manifolds and $f: A \to S$ a C^∞-embedding, then for any C^∞-manifold M the induced C^∞-map $\overline{f}: C^r(S,M) \to C^r(A,M)$ is a foliation map over its image $(0 \leq r < \infty)$. We then obtain [25] the

THEOREM. $\overline{f}: C^r(S,M) \to C^r(A,M)$ *is a locally* C^0*-trivial fibration over its image.*

The space $Em^r(S,M)$ of C^r-embeddings of S in M is an open submanifold of $C^r(S,M)$. The following result is due to Thom [29], Cerf [3], and Palais [25]; see also [20]:

THEOREM. $\overline{f}: Em^r(S,M) \to Em^r(A,M)$ *is a locally* C^0*-trivial fibration over its image* $(2 \leq r \leq \infty)$.

We can suppose that S and A have boundaries; and S need not be compact.

Analogously [4], *let* $\tilde{\mathfrak{D}}$ *be an open subgroup of* $\mathfrak{D}(S)$. *Then* $\tilde{\mathfrak{D}}$ *operates principally on* $Em^r(S,M)$.

EXAMPLE. Taken together with Example 4B, we find the following result, of interest in the calculus of variations: *If* S *is a closed surface of genus* (S) ≥ 2, *then the orbit map* $\mathrm{Em}^r(S,M) \to \mathrm{Em}^r(S,M)/\mathfrak{D}_0(S)$ *is a homotopy equivalence.*

(C) We fix an embedding $h: S \to M$ and thereby view S as a submanifold of M. Denote by $\mathrm{Em}^r(S,M;A)$ the totality of C^r-embeddings of S into M which induce the identity on A. Let $J_A^r\mathrm{Em}^r(S,M;A)$ denote the space of r-jets of these embeddings which are tangent through order r at every point of $A (1 \leq r \leq \infty)$.

The following result – and its many variants – play a fundamental role in the theory of Cerf [3, 4]:

THEOREM. *The canonical map*

$$\mathrm{Em}^r(S,M;A) \to J_A^r\mathrm{Em}^r(S,M;A)$$

is a locally trivial fibration.

EXAMPLE. If M is a compact m-manifold without boundary and D^p the closed p-dimensional Euclidean disc centered at 0 $(p \leq m)$, then

$$\bar{f}: \mathrm{Em}^\infty(D^p,M) \to J_0^1\mathrm{Em}^\infty(D^p,M;0)$$

is a locally trivial fibration. Now we have a canonical identification of $J_0^1\mathrm{Em}^\infty(D^p,M;0)$ with the Stiefel manifold $V_{m,p}(M)$ of p-frames of M. Furthermore, $\bar{f}$ has aspherical fibres, whence *we have a homotopy equivalence* $\bar{f}: \mathrm{Em}^\infty(D^p,M) \to V_{m,p}(M)$.

(D) A corresponding theory of fibrations for immersions is more delicate and difficult.

The space $\mathrm{Im}^r(S,M)$ of C^r-immersions of S in M is also an open submanifold of $C^r(S,M)$. The following result was first established (in a slightly different form) by Smale and Thom ([29]; see [26] for further bibliography), with refinements made by Hirsch-Palais.

THEOREM. *If* $\dim S < \dim M$, *then the restriction map* $f: \mathrm{Im}^r(S,M) \to \mathrm{Im}^r(A,M)$ *is a Hurewicz fibration over its image* $(2 \leq r \leq \infty)$.

Again, we can suppose that S and A have boundaries.

A primary application of that fibration is the *fundamental classification theorem for immersions of* Hirsch-Smale [26]:

If dim S < dim M, *then the map* $\phi \to \phi_*$ *(the differential of* ϕ*) induces a bijective correspondence between the regular homotopy classes of immersions of* S *in* M *and the (monomorphism) homotopy classes of monomorphisms of* T(S) *into* T(M).

BIBLIOGRAPHY

[1] R. Arens, *Topologies for homeomorphism groups.* Amer. J. Math. 68 (1946), 593-610.

[2] H. Cartan, Sém. E. N.S. 1948/9.

[3] J. Cerf, *Topologie de certains espaces de prolongements.* Bull. Soc. math. France, 89 (1961), 227-382.

[4] __________, *Théorèmes de fibration* .... Sém. Cartan E.N.S. 1962/3. Exp. 8.

[5] J. Dixmier, Les C*-algèbres et leur représentations. Gauthier-Villars 1964.

[6] A. Dold, *Partitions of unity in the theory of fibrations.* Annals of Math. 78 (1963), 223-255.

[7] A. Douady, *Le probleme des modules* .... Ann. Inst. Fourier, Grenoble, 16 (1) (1966), 1-98.

[8] C. J. Earle and J. Eells, *On the differential geometry of Teichmüller spaces.* J. d'Analyse.

[9] __________, *Foliations and fibrations.* J. Differential Geometry 1 (1967).

[10] __________, *The diffeomorphism group of a compact Riemann surface.* Bull. A.M.S. (1967).

[11] J. Eells, *A setting for global analysis.* Bull. A.M.S. 72 (1966), 751-807.

[12] C. Ehresmann, *Les connexions infinitésimales dans un espace fibré différentiable.* Colloque de Topologie, Bruxelles (1950), 29-55.

[13] R. H. Fox, *On fibre spaces II.* Bull. A.M.S. 49 (1943), 733-735.

[14] A. Grothendieck, *Espaces vectoriels topologiques.* São Paulo, 1964.

[15] M-E. Hamstrom, *Homotopy groups of the space of homeomorphisms on a 2-manifold.* Ill. J. Math. 10 (1966), 563-573.

[16] R. Hermann, *A sufficient condition that a mapping of Riemannian manifolds be a fibre bundle.* Proc. A.M.S. 11 (1960), 236-242.

[17] S-T. Hu, Homotopy theory. Academic Press, 1959.

[18] W. Hurewicz, *On the concept of fiber space.* Proc. N.A.S. 41 (1955), 956-961.

[19] J. A. Leslie, *On a differential structure for the group of diffeomorphisms.* Topology (to appear).

[20] E. L. Lima, *On the local triviality of the restriction map for embeddings.* Comm. Math. Helv. 38 (1963/64), 163-164.

[21] G. S. McCarty, *Homeotopy groups.* Trans. A.M.S. (1963), 293-304.

[22] E. Michael, *Convex structures and continuous selections.* Can. J. Math. 11 (1959), 556-575.

[23] J. Milnor, *Construction of universal bundles II.* Annals of Math. 63 (1956), 430-436.

[24] J. C. Moore, *On a theorem of Borsuk.* Fund. Math. 43 (1956), 195-201.

[25] R. S. Palais, *Local triviality of the restriction map for embeddings.* Comm. Math. Helv. 34 (1960), 305-312.

[26] S. Smale, *A survey of some recent developments in differential topology.* Bull. A.M.S. 69 (1963), 131-145.

[27] N. E. Steenrod, The topology of fibre bundles. Princeton (1951).

[28] R. Thom, *Opérations en cohomologie réelle.* Sém. H. Cartan E.N.S. (1954/5), Exp. 17.

[29] ____________, *La classification des immersions.* Sém. Bourbaki (1957/8), Exp. 157.

CORNELL UNIVERSITY;
CHURCHILL COLLEGE, CAMBRIDGE.

AN APPROXIMATE MORSE-SARD THEOREM

by

JAMES EELLS[1] AND JOHN MCALPIN[2]

Let X and Y be smooth manifolds, and let $\phi: X \to Y$ be a smooth map. Say that $x \in X$ is a critical point of ϕ if its differential $\phi_*(x): X(x) \to Y(\phi(x))$ is not surjective. The set C_ϕ of all critical points is closed in X. Its image $\phi(C_\phi)$ is the *set of critical values of* ϕ, and we will say that ϕ is a *Sard map* if $\phi(C_\phi)$ has no interior point in Y. We will say that ϕ is a μ-*Sard map*—relative to some measure μ on Y—if $\mu(\phi((C_\phi)) = 0$.

A theorem of primary importance in differential topology is the following (14, Ch. I): *Any* C^r-*map* $\phi: X \to Y$ *is a* μ_m-*Sard map, provided that* $r > \max(n-m, 0)$, *where* $n = \dim X$, $m = \dim Y$, *and* μ_m *is a Lebesgue* m-*measure on* Y.

We now consider the possibility of obtaining a theorem of that sort for manifolds modelled on Banach spaces. There have been some results in this area—notably that of Smale [13] to the effect that Fredholm maps between smooth manifolds, modelled on separable Banach spaces, are Sard maps. Such results have been highly restrictive; on the other hand, examples, such as that of Kupka [8], show that strong restrictions are necessary.

For many applications, however, it will suffice to know that a given map can be approximated (in a suitable sense) by a Sard map. In that di-

---

1 Research partially supported by NSF Grant GP-4216.

2 Research sponsored by the Air Force Office of Scientific Research, Office of Aerospace Research, United States Air Force, under AFOSR Grant Nr. 1243-67.

rection we have established the

THEOREM: *Let* X *and* Y *be smooth manifolds modelled on a Hilbert space* E *and a Banach space* F, *respectively. Suppose that* X *is* separable. *Then the Sard maps are dense in the fine topology on* $C^0(X, Y)$.

The proof is reduced to a local situation by standard techniques. The approximation itself is achieved by using a smooth partition of unity, whose summands are specially constructed functions (using the method of scalloping [9]) with critical points that can be kept under control—so that we can use the Morse-Sard theorem.

The above theorem permits us to extend a standard transversality theorem as follows:

THEOREM: *Let* X *and* Y *be smooth manifolds as in the preceding theorem. Let* B *be a closed direct submanifold. Then for any smooth map* $\phi: X \to Y$ *and any smooth function* $\varepsilon: X \to R$ (> 0) *there is an* ε*-approximation* $\psi: X \to Y$ *of* ϕ *which is transversal over* B.

A smooth submanifold B of Y is *direct* if every tangent subspace B(y) is a direct summand of Y(y).

As an application (of a special case) we can show that *every closed subset of a smooth Hilbert manifold has a fundamental system of smooth neighborhoods.* (An open neighborhood is called smooth if its boundary is a smooth 1-codimensional submanifold.) This should be viewed in conjunction with the basic separation theorem [5, p. 412] for closed convex subsets of a Hilbert space.

We are also able to establish a form of the Atiyah-Thom duality Theorem for Hilbert manifolds.

CORNELL UNIV.
CHURCHILL COLLEGE, UNIV. OF COLORADO

BIBLIOGRAPHY

[1] R. Abraham and J. Robbin, *Transversal Mappings and Flows*, Benjamin (1967).

[2] M. Atiyah,"Bordism and Cobrodism," *Proc. Camb. Phil. Soc.*, *57* (1961), 200-208.

[3] R. Bonic and J. Frampton, "Smooth functions on Banach Manifolds." *J. Math. Mech.* *15* (1966), 877-898.

[4] P. Conner and E. Floyd, "Differentiable periodic maps," *Erget. Math. Bd.*, *33* (1964).

[5] J. Dunford and J. Schwartz, *Linear Operators*, Interscience, 1958.

[6] J. Eells, "A setting for global analysis, *Bull. Am. Math. Soc.* *72* (1966), 751-807.

[7] E. Feldman, "The Geometry of Immersions I, *TAMS*, *120* (1966), 185-224.

[8] I. Kupka, "Counterexample to the Morse-Sard theorem in the case of Infinite Dimensional Manifolds," *Proc. A. M. S.*, *16* (1965), 954-F.

[9] S. Lang, *Introduction to Differentiable Manifolds*, Interscience, 1962.

[10] A. Sard, "The measure of the critical values of differentiable maps," *Bull. A. M. S.*, *48* (1942), 883-890.

[11] _____, "Images of critical sets," *Annals of Math.*, *68* (1958), 247-259.

[12] _____, "Hausdorff measure of critical images on Banach manifolds," *Am. J. Math.*, *87* (1965), 158-174.

[13] S. Smale, "An infinite dimensional version of Sard's theorem," *Am. J. Math.*, *87* (1965), 861-866.

[14] R. Thom, "Quelques proprietes globales des varites differentiables," *Comm. Math. Helv.*, *28* (1954), 17-86.

[15] H. Whitney, "A function not constant on a connected set of critical points," *Duke Math. J.*, *1* (1935), 514-517.

MORSE THEORY FOR CLOSED CURVES

HALLDOR I. ELIASSON

Introduction.

Let X be a complete Riemannian manifold of class C^∞, modeled on separable Hilbert spaces and let F be a C^∞ function on X. In order to provide Morse theory for F, we must prove: F satisfies condition (C) of Palais and Smale [9] and F has only nondegenerate critical manifolds Bott [1], Meyer [10], Wasserman [12].

We will here discuss the energy function defined on a Hilbert manifold of closed curves and present some intrinsic methods, which are both practical and extendable to more general variational problems. The main difference between the case of closed curves and curves connecting two fixed points is, that we have necessarily degenerate critical points in the first case (critical manifolds of $\dim \geq 1$). Several people have observed, that the proof of Palais [8] for condition (C) in the second case can also be used for closed curves; we will, however, give an independent completely intrinsic proof here (§3). We will first outline the theory and then present the proofs and the tools from differential geometry used. A more general and complete introduction to the differential geometric aspect can be found in [5].

§1. *Statement of results.*

In what follows M is a compact Riemannian manifold of class C^∞, connected and without boundary. $\tau\colon TM \to M$ will denote its tangent bundle and $\exp\colon TM \to M$ the exponential map. $S = \mathbf{R}/\mathbf{Z}$ is the 1-torus (or circle). We will use H^0 and H^1 to denote the class of maps, which are square integrable and have square integrable derivatives respectively. $H^1(S, M)$ the set of maps $x: S \to M$ of class H^1 is a Hilbert manifold of class C^∞. The tangent space at x can be identified with the Hilbert space of H^1-fields $\xi: S \to TM$ along x, $\tau \circ \xi = x$, or equivalently with $H^1(x^*\tau)$, sections of class H^1 in the pull-back $x^*\tau$ of the tangent bundle τ of M by x. $\xi \to \exp \circ \xi$ can be interpreted as the exponential map for $H^1(S, M)$ given by a certain connection for this manifold [5] and provides it in particular with charts. A Riemannian metric can be defined in $H^1(S, M)$ by:

$$\begin{aligned} <\xi, \eta>_1 &= \int_0^1 <\xi(t), \eta(t)>dt + \int_0^1 <\nabla \xi(t), \nabla \eta(t)>dt \\ &= <\xi, \eta>_0 + <\nabla \xi, \nabla \eta>_0 , \end{aligned}$$

where $< \, , \, >$ denotes the Riemannian metric on M and Δ the corresponding covariant differentiation.

THEOREM 1. $H^1(S, M)$ *is a complete Riemannian manifold of class* C^∞, *modeled on separable Hilbert spaces.*

The energy function

$$E(x) = \frac{1}{2} \int_0^1 \|\partial x\|^2 dt = \frac{1}{2} \|\partial x\|_0^2$$

is now a C^∞ function $E: H^1(S, M) \to \mathbf{R}$ and its derivative, as a section in the cotangent bundle, is

$$dE(x) \cdot \eta = <\partial x, \eta>_0, \quad \eta \in H^1(x^*TM) .$$

Suppose x is a geodesic, i.e., x is of class C^∞ and $\nabla \partial x = 0$, then $dE(x) \cdot \eta = -<\nabla \partial x, \eta>_0 = 0$, thus x is a critical point of E, $dE(x) = 0$. Conversely, if $x \in H^1(S, M)$ and $dE(x) = 0$, then $<\partial x, \nabla \eta>_0 = 0$ for all $\eta \in H^1(x^*TM)$ is easily seen to imply, that x is of class C^∞ (regularity) and thus $\nabla \partial x = 0$.

THEOREM 2. *The critical points of* E *are exactly the closed geodesics in* M.

THEOREM 3. E *satisfies condition* (C)*: Given any sequence* x_k *in* $H^1(S, M)$*, such that* $E(x_k)$ *is bounded and* $\|dE(x_k)\|$ *converges to zero, then* x_k *has a convergent subsequence, converging to a critical point of* E.

Let $R: TM \oplus TM \oplus TM \to TM$ be the curvature tensor on M, put $\nabla^2 = \Delta$ (Laplacian) and define $K_x: H^1(x^*\tau) \to H^1(x^*\tau)$ by

$$K_x \cdot \xi = R \circ (\partial x, \xi, \partial x) \quad \text{for } x \in C^\infty(S, M) \ .$$

THEOREM 4. *The Hessian of* E *at a critical point* x *of* E *is given by:*

$$H(E)_x \cdot (\xi, \eta) = \langle \nabla \xi, \nabla \eta \rangle_0 + \langle \partial x, R \circ (\xi, \partial x, \eta) \rangle_0$$
$$= \langle A_x \cdot \xi, \eta \rangle_1$$

where $A_x: H^1(x^*\tau) \to H^1(x^*\tau)$ *is the self-adjoint Fredholm operator:*

$$A_x = 1 + (1 - \Delta)^{-1} \circ (K_x - 1) \ .$$

COROLLARY 1. *If* x *is a critical point of* E*, then we have an orthogonal decomposition:*

$$H^1(x^*\tau) = T_x^0 + T_x^- + T_x^+$$

is the sum of eigenspaces of A_x *corresponding to zero, negative and positive eigenvalues* λ. *Moreover*

$$T_x^0 = \ker(-\Delta + K_x) \subset C^\infty(x^*\tau)$$

$$T_x^- = \Sigma_{\lambda < 0} \ker((\lambda - 1)\Delta + K_x - \lambda) \subset C^\infty(x^*\tau)$$

and

$$\text{Nullity}(x) = \dim T_x^0 < \infty$$
$$\text{Index}(x) = \dim T_x^- < \infty \ .$$

Proof: $(\lambda-1)\Delta + K_x - \lambda$ is an elliptic differential operator and there are only finitely many negative eigenvalues λ of A_x.

Definition. Given a C^∞ curve c in M, a C^∞ field α along c is called a Jacobi field, iff it satisfies the differential equation:

$$\nabla^2 \alpha + R \circ (\alpha, \partial c, \partial c) = (\Delta - K_c) \cdot \alpha = 0 \ .$$

We will say, that M has property (J), iff given a closed geodesic $c: S \to M$ and a closed Jacobi field $\xi: S \to TM$ along c, then there is an infinitesimal isometry Y on M, such that $\xi = Y \circ c$.

COROLLARY 2. M *imbedded in* $H^1(S, M)$ *as a submanifold (point curves) is a nondegenerate critical manifold of index zero of the energy function.*

Proof: The imbedding $i: M \to H^1(S, M)$ is given by $i(p)(t) = p$, $t \in S$. Then $Ti(v)(t) = v$ for all $v \in TM$ and obviously the Jacobi fields along $x = i(p)$ are exactly the constant maps into T_pM, i.e., T_x^0 is the tangent space to $i(M)$ at $i(p) = x$. Moreover (observe that $\partial x = 0$ implies $K_x = 0$):

$$[(\lambda-1)\Delta + K_x - \lambda] \cdot \xi = 0$$

has no periodic solutions if $\lambda < 0$, as it reduces to the ordinary differential equation:

$$\xi'' = \frac{-\lambda}{1-\lambda}\, \xi \quad \text{in} \quad T_pM \ .$$

COROLLARY 3. *Zero is an isolated critical value of* E *and there is a smallest positive critical value, establishing the existence of a closed geodesic of minimal length on* M.

Proof: If there are no positive critical values, then M is contained as a deformation retract of $H^1(S, M)$, which is absurd, as then the loop space of M is homotopically trivial, but not M. Using Theorem 3, the set of critical points with bounded energy is compact, thus the lower bound for positive critical values is taken and must be positive by Corollary 2 and Morse lemma (nondegenerate critical levels are isolated). The existence of closed nontrivial geodesics on even a simply connected manifold is a result obtained by Fet [6] and others, obviously we have one in any nontrivial homotopy class of $\pi_1(M)$ as a minimum point of E in the corresponding component of $H^1(S, M)$. Let $I(M)$ be the group of isometries of M. $I(M)$ is a compact Lie transformation group of M and acts on $H^1(S, M)$ by composition $(g, x) \to g \circ x$. This map $I(M) \times H^1(S, M) \to H^1(S, M)$ is of class C^∞ and its tangent at (g, x) is

$$(Y, \xi) \to Y \circ g \circ x + (dg \circ x) \cdot \xi ,$$

where Y is an infinitesimal isometry and $\xi \in H^1(x^*\tau)$. In particular $I(M)$ is an effective Lie transformation group of $H^1(S, M)$. The orbits are compact submanifolds and the tangent bundle of any orbit is obtained by restricting the infinitesimal isometries to the curves in the orbit. The energy function is obviously invariant under this action, so every orbit $I(M) \cdot c$, c a closed geodesic, is a critical manifold of E.

THEOREM 5. *Suppose* M *has property* (J), *then* E *has only nondegenerate critical manifolds.*

Proof: Property (J) implies that T_c^0 is exactly the tangent space at c to the orbit $I(M) \cdot c$ and as A_c is a linear homeomorphism of $T_c^- + T_c^+$, the Hessian is nondegenerate on $T_c^- + T_c^+$.

My conjecture is, that every (irreducible) globally Riemannian symmetric space has property (J). This is a similar condition as "variational completeness" required by Bott and Samelson [2] in their study of Morse theory for loop spaces. Their proof of variational completeness in the symmetric case, implies (J) for those Jacobi fields vanishing at some point of the curve, however more is needed if the rank is larger than 1.

Example. $M = S^n$ the unit sphere in $\mathbf{R}^{n+1}$.

For every nonnegative integer j we have a nondegenerate critical manifold W_j of the energy function. W_j is the submanifold of closed geodesics of length $2\pi j$. We can identify W_0 with S^n and W_j with the Stiefel manifold of 2-frames in $\mathbf{R}^{n+1}$:

$$W_j = V_{n+1,2} = SO_{n+1}/SO_{n-1}, \quad j \geq 1 .$$

We can easily solve the differential equation involved (Corollary 1) to compute the index $k_j = \text{Index}(c)$ for $c \in W_j$. We obtain $k_0 = 0$ and

$$k_j = (2j-1)(n-1) \quad \text{for } j \geq 1 .$$

Let a_j be a sequence of numbers with

$$E(W_j) < a_j < E(W_{j+1}), \quad j \geq 0$$

and put $X_j = E^{-1}([0, a_j])$, $j \geq 0$. If H_* denotes the singular homology functor, we obtain from Morse theory ([10] and [12]) using any group G of coefficients:

$$H_k(X_j, X_{j-1}; G) = H_{k-k_j}(W_j; G) \qquad j \geq 0 ,$$

for the homology group in dimension k.

THEOREM 6. *For* $n > 2$ *we have for all* $k \geq 0$:

$$H_k(H^1(S, S^n); G) = H_k(S^n; G) + \sum_{j=1}^{\infty} H_{k-(2j-1)(n-1)}(V_{n+1,2}; G) .$$

Proof: We prove $H_k(X_j) = \Sigma_{i=0}^{j} H_k(X_i, X_{i-1})$ using induction on j. As this formula is trivial for $j = 0$, suppose it is true for j and consider the exact sequence

$$\cdots \to H_{k+1}(X_{j+1}, X_j) \to H_k(X_j) \to H_k(X_{j+1}) \to$$
$$\to H_k(X_{j+1}, X_j) \to H_{k-1}(X_j) \to \cdots \quad .$$

Now the first non-vanishing homotopy group of $V_{n+1,2}$ in positive dimension is $\pi_{n-1}(V_{n+1,2}) = Z$ if n is odd and $= Z_2$ if n is even (see [13]), therefore $V_{n+1,2}$ has nonzero homology at most in the dimensions 0, n − 1, n, 2n − 1. Then using $k_{j+1} = k_j + 2n - 2$ and $n > 2$ implies $2n-2 > n$, $2n-2+n-1 > 2n-1$ we see that if $H_k(X_{j+1}, X_j) \neq 0$ then $H_k(X_j) = H_{k-1}(X_j) = 0$ and if $H_k(X_j) \neq 0$ then $H_k(X_{j+1}, X_j) = H_{k+1}(X_{j+1}, X_j) = 0$ and so for all k: $H_k(X_{j+1}) = H_k(X_{j+1}, X_j) + H_k(X_j)$, which proves the theorem.

Using Z_2 coefficients, we obtain for the Poincaré polynomial (or series):

$$P(H^1(S, S^n); z) = P(S^n; z) + P(V_{n+1,2}; z)z^{n-1}(1-z^{2(n-1)})^{-1} .$$

This formula and a comparison with the polynomial having the circular connectivities or sensed circular connectivities of S^n as coefficients (see Bott [1]), suggests that those are exactly the mod 2 Betti numbers of the spaces $H^1(S, S^n)/O_2$ and $H^1(S, S^n)/SO_2$ respectively (where the action of O_2 is the unusal rotation of the closed curves).

Using the information $H_k(V_{n+1,2}; Z) = Z$ if $k = 0, 2n-1$ or $n-1$, n and n odd and $= Z_2$ in case $k = n-1$ and n even and $= 0$ elsewhere (as follows from our knowledge of π_{n-1}, using Poincaré duality and the universal coefficient theorem), we obtain for the integer homology:

If n is odd, then

$$H_k(H^1(S, S^n)) = \begin{cases} Z & \text{for } k = j(n-1) \text{ and } j(n-1)+n \text{ for all } j \geq 0 \\ 0 & \text{otherwise.} \end{cases}$$

If n is even, then

$$H_k(H^1(S, S^n)) = \begin{cases} Z & \text{for } k = 0,\ (2j+1)(n-1)+1 \\ & \quad \text{and } (2j+1)(n-1) \text{ for all } j \geq 0 \\ Z_2 & \text{for } k = 2j(n-1) \text{ all } j \geq 1 \\ 0 & \text{otherwise.} \end{cases}$$

This result agrees with Švare [11] and Eells [4], except in [4] the torsion (n even) is missing. This however, is due to an incorrect statement in Lemma (b) page 121 [4], the homomorphism $j_n\colon H_n(X) \to H_n(X, Y^*)$ is not an isomorphism for n even, but rather a multiplication by 2: $z \to Z$, thus $H^n(Y^*) = Z_2$ and not 0. Eells uses the Banach manifold $C^0(S, S^n)$ of continuous maps and we know from a general theorem of Palais [7], that the continuous inclusion: $H^1(S, M) \subset C^0(S, M)$ is a homotopy equivalence.

§2. *The manifold* $H^1(S, M)$.

As before $\tau\colon TM \to M$ is the tangent bundle and $\exp\colon TM \to M$ the exponential map. Let $D_2 \exp\colon\ TM \to L(TM, \exp{}^*TM)$ denote the fibre derivative of exp:

$$D_2 \exp(v) \cdot u = \frac{d}{dt} \exp(v + tu) \,|\, t = 0 \;.$$

Let $\mathcal{O}$ be an open neighborhood of the zero section in TM, such that $(\tau, \exp)$ maps $\mathcal{O}$ diffeomorphic onto an open neighborhood of the diagonal in $M \times M$. Then for $v \in \mathcal{O}_p$

$$D_2 \exp(v)\colon\ T_pM \to T_{\exp v}M$$

is a linear isomorphism.

Given a C^∞ map $c\colon S \to M$, $\mathcal{O}_c = c^*\mathcal{O}$ is an open neighborhood of the zero section in the pull-back c^*TM, and the correspondence

$$C^0(\mathcal{O}_c) \ni \xi \longleftrightarrow x = \exp \circ \xi \in C^0(S, M)$$

is one-to-one. Moreover the transition $\exp \circ \xi_1 = \exp \circ \xi_2$ between sections in $\mathcal{O}_{c1}$ and $\mathcal{O}_{c2}$ for two close C^∞ maps c_1, c_2 is given by $\xi_2 = F \circ \xi_1$, where F is a fibre-preserving map of class C^∞. Now this holds for any compact C^∞ manifold S and the main idea in the construction of manifolds of maps from S to M, Eells [3], is to use this 1:1 correspondence as a chart, that is by restricting it to the Banach space of sections in pull-backs by C^∞ maps $c\colon S \to M$ used as model. A class of Banach spaces of sections serving this purpose has been axiomatically described by Palais [7] and also in [5] by the author. This class includes the C^k, $0 \leq k < \infty$, spaces and the Sobolev H^k spaces if $k > \frac{1}{2} \dim S$. We will here only be interested in H^0, H^1 and the case, where S is the circle.

Let E and F be vector bundles over S, say pull-backs by C^∞ maps from S into M, and give them the Riemannian metric and connection induced from the tangent bundle of M. We then have inner products in $H^0(E)$ and $H^1(E)$ by

$$<\xi, \eta>_0 = \int_0^1 <\xi(t), \eta(t)> dt$$

$$<\xi, \eta>_1 = <\xi, \eta>_0 + <\nabla\xi, \nabla\eta>_0 \quad ,$$

we denote the corresponding norms by $\| \ \|_0$ and $\| \ \|_1$ and define the usual norm $\| \ \|_\infty$ in $C^0(E)$ by $\|\xi\|_\infty = \sup \|\xi(t)\|$ $(t \in S)$.

We then have the following properties:

Property 1. We have continuous linear inclusions

$$H^1(E) \subset C^0(E) \subset H^0(E) ,$$

where the first inclusion is completely continuous.

In fact: $\|\xi\|_0 \leq \|\xi\|_\infty \leq 2\|\xi\|_1$.

Property 2. We have a continuous linear inclusion

$$H^1(L(E, F)) \subset L(H^\nu(E), H^\nu(F)) \quad \nu = 0, 1 .$$

In fact: $\|A \cdot \xi\|_\nu \leq 2\|A\|_1 \cdot \|\xi\|_\nu$.

Property 3. Let $\mathcal{O}$ be an open subset of E projected onto S and let $f: \mathcal{O} \to F$ be a fibre map of class C^∞, then $\tilde{f}(\xi) = f \circ \xi \in H^1(F)$ whenever $\xi \in H^1(\mathcal{O})$ and the map $\tilde{f}: H^1(\mathcal{O}) \to H^1(F)$ is continuous.

From these three properties one obtains easily the following fundamental lemma (see [5], §4).

LEMMA. $\tilde{f}: H^1(\mathcal{O}) \to H^1(F)$ *is of class* C^∞ *and its derivative is the map:*

$$Df^{\sim} = (D_2 f)^{\sim}\colon H^1(\mathcal{O}) \to H^1(L(E,F)) \subset L(H^1(E), H^1(F)),$$

where $D_2 f\colon \mathcal{O} \to L(E,F)$ *is the fibre derivative of* f, *i.e.*, $D_2 f \mid \mathcal{O}_t = D(f \mid \mathcal{O}_t)$ for $t \in S$.

Then using the results described in §5 and 6 in [5] we obtain the following:

(a) $H^1(S,M)$ is a manifold of class C^∞, modeled on the separable Hilbert spaces $H^1(c^*TM)$ with $c \in C^\infty(S,M)$ and the map

$$\exp^{\sim}\colon H^1(\mathcal{O}_c) \to H^1(S,M)$$

provides a chart, the natural chart centered at c.

(b) We have C^∞ vector bundles

$$H^\nu(H^1(S,M)^*TM) \to H^1(S,M) \quad \nu = 0, 1$$

with $H^\nu(x^*TM)$ as the fibre over $x \in H^1(S,M)$. A local trivialization over the natural chart at c is given by:

$$(D_2 \exp)^{\sim}\colon H^1(\mathcal{O}_c) \times H^\nu(c^*TM) \to H^\nu(H^1(S,M)^*TM) .$$

For $\nu = 1$ this bundle is naturally equivalent with the tangent bundle of $H^1(S,M)$ and the local trivialization above corresponds to the tangent trivialization $(\exp^{\sim}.\ T\exp^{\sim})$ of the tangent bundle under the equivalence (if we put $a(s,t) = \exp(\xi(s) + t\eta(s))$ with $\xi, \eta \in H^1(x^*\tau)$, then $\partial_2 a(s,t) = D_2 \exp(\xi(s) + t\eta(s)) \cdot \eta(s)$ gives the tangent field of $a^{\sim}$, $a^{\sim}(s)(t) = a(t,s)$, by $\partial a^{\sim} = (\partial_2 a)^{\sim}$). Moreover the total space is identical with the manifold $H^1(S,TM)$ if we identify fields $\xi\colon S \to TM$ along a map $x = \tau \circ \xi\colon S \to M$ with the sections in the pull-back by x.

(c) Taking the tangent field of curves in $H^1(S,M)$ as a C^∞ section ∂ in the bundle

$$H^0(H^1(S,M)^*TM) .$$

In the local trivialization centered at c, the local principal part of ∂:

$$\partial_c : H^1(\mathcal{O}_c) \to H^0(c^*TM)$$

is given by:

$$\partial_c(\xi) = \nabla\xi + \theta_c \circ \xi, \text{ with } \theta_c(v_s) = \theta(v) \cdot \partial c(s)$$

and where $\theta: \mathcal{O} \to L(TM, TM)$, $\mathcal{O} \subset TM$, is the map

$$\theta(v) \cdot u = [D_2 \exp(v)]^{-1} \cdot \nabla_1 \exp(v, u) \ .$$

(d) We have a unique continuous extension of the covariant differentiation of fields in TM, to a C^∞ section Δ in

$$L(H^1(H^1(S, M)^* TM), H^0(H^1(S, M)^* TM))$$

with the local principal part:

$$\nabla_c(\xi) \cdot \eta = \nabla\eta + (D_2\theta_c \circ \xi) \cdot \eta + (\Lambda_c \circ \xi) \cdot (\partial_c(\xi), \eta) \ .$$

Here $\Lambda: \mathcal{O} \to L_s^2(TM, TM)$ is given by

$$\Lambda(v) \cdot (u, w) = [D_2 \exp(v)]^{-1} \cdot D_2^2 \exp(v) \cdot (u, w)$$

and $\Lambda_c = c^*\Lambda$.

(e) The inner products $< \ , \ >_\nu$ in the model spaces $H^\nu(c^*TM)$, $c \in C^\infty(S, M)$, have a unique extension to a Riemannian metric in the vector bundles $H^\nu(H^1(S, M)^* TM)$ $\nu = 0, 1$, of class C^∞. In particular for $\nu = 1$, we obtain a Riemannian metric for $H^1(S, M)$.

Define a map $G: \mathcal{O} \to L(TM, TM)$ by

$$<D_2 \exp(v) \cdot u, D_2 \exp(v) \cdot w> = <u, G(v) \cdot w> ,$$

$G(v)$ is then a positive self-adjoint operator and G is a fibre map of class C^∞. In the local trivialization centered at c, $< \ , \ >_0$ is given by

$$<(G_c \circ \xi) \cdot \eta, \zeta>_0, \ G_c = c^*G; \ \xi \in H^1(\mathcal{O}_c); \ \eta, \zeta \in H^0(c^*\tau)$$

at the point $x = \exp \circ \xi$. Using the fundamental lemma and Property 2,

$$G_c^{\sim}: H^1(\mathcal{O}_c) \to H^1(L(c^*TM, c^*TM)) \subset L(H^0(c^*TM), H^0(c^*TM))$$

is of class C^∞ and similarly $<\ ,\ >_1$ is of class C^∞ using (d).

(f) We have a Riemannian connection for the vector bundle $H^0(H^1(S,M)^*TM)$ given by the local connector

$$\Lambda_c^{\sim}: H^1(\mathcal{O}_c) \to H^1(L_s^2(E,E)) \subset L(H^0(E), H^1(E); H^0(E)) ,$$

where $E = c^*TM$ and Λ_c is the map defined in (d). In this connection ∇ is the covariant derivative of the section ∂ as follows immediately from the local formulas in (c) and (d), i.e.,

$$D\partial_c(\xi)\cdot\eta + \Lambda_c^{\sim}(\xi)(\partial_c(\xi),\eta) = \nabla_c(\xi)\cdot\eta .$$

In order to prove that Λ gives the local connector for a Riemannian connection, we must show:

$$2<G(v)\cdot\Lambda(v)\cdot(\xi,\eta),\zeta> = <D_2G(v)\cdot(\xi,\nu),\zeta> +$$

$$+ <D_2G(v)\cdot(\eta,\zeta),\xi> - <D_2G(v)\cdot(\zeta,\xi),\eta> ,$$

but this follows easily by differentiating both sides of the equation in (e) defining G. As a matter of fact, this Riemannian connection is unique as H^1 is dense in H^0.

We have here proved all the ingredients of Theorem 1 except for completeness. By Property 1 a Cauchy sequence in $H^1(S,M)$ must be a Cauchy sequence in $C^0(S,M)$, then using the obvious completeness of $C^0(S,M)$, we can find a $c \in C^\infty(S,M)$ so close to the limit in $C^0(S,M)$ that all but finitely many elements of the sequence are in the domain of the natural chart centered at c, then we obtain convergence in $H^1(S,M)$ by completeness of the model $H^1(c^*TM)$.

§3. *The energy function.*

The energy function $E: H^1(S,M) \to \mathbf{R}$

$$E(x) = \tfrac{1}{2}\|\partial x\|_0^2$$

is of class C^∞ as the composite of the C^∞ section ∂ in $H^0(H^1(S,M)^*TM)$ and 1/2 times the square of the norm in this bundle. As Δ is the covariant derivative of ∂ in the Riemannian connection (see §2(f)) the derivative is:

$$dE(x)\cdot\eta = \langle\partial x, \nabla\eta\rangle_0, \qquad \eta \in H^1(x^*\tau) ,$$

where we write $\Delta\eta$ short for $\Delta(x)\cdot\eta$. The local representative of E in the natural chart centered at c is given by:

$$E_c(\xi) = \tfrac{1}{2}\langle(G_c\circ\xi)\cdot\partial_c(\xi), \partial_c(\xi)\rangle_0 ,$$

using the local formulas for ∂ and $\langle\ ,\ \rangle_0$ given in §2. Similarly using (d) and differentiating E_c:

$$DE_c(\xi)\cdot\eta = \langle(G_c\circ\xi)\cdot\partial_c(\xi), \nabla_c(\xi)\cdot\eta\rangle_0$$

$$= \langle(G_c\circ\xi)\cdot\partial_c(\xi), \Delta\eta + (D_2\theta_c\circ\xi)\cdot\eta\rangle_0$$

$$+ \tfrac{1}{2}\langle(D_2G_c\circ\xi)\cdot(\eta,\partial_c(\xi)), \partial_c(\xi)\rangle_0 .$$

We can check the second equality by using the relation between D_2G_c and Λ_c from (f), §2.

We will now turn to the proof of Theorem 3 and prove first:

For any number $a < \infty$, $E^{-1}([0,a])$ is a relatively compact subset of $C^0(S,M)$.

If d_M denotes the distance function in M, then for any $x \in H^1(S,M)$ we have

$$d_M(x(t), x(s)) \le \left|\int_t^s \|x(r)\|\,dr\right| \le \sqrt{|s-t|}\cdot\sqrt{2E(x)} ,$$

using Hölder inequality. Thus $E^{-1}([0,a)]$ is an equicontinuous subset of $C^0(S,M)$ and is then relatively compact by Ascoli's theorem, as M is compact.

Thus given a sequence $x_n \in H^1(S,M)$, $n = 1, 2, \dots$ with $E(x_n) \le a$, we may assume that x_n converges to some continuous map x_0 in the uniform topology. We then choose a $c \in C^\infty(S,M)$ so close to x_0, that x_0 is well in the domain of the chart for $C^0(S,M)$ at c, let then ξ_i, $i = 1, 2, \ldots$ be the sequence in $H^1(c^*\tau)$ corresponding to x_n, we have $\|\xi_i - \xi_0\| \to 0$, where $\exp \circ \xi_0 = x_0$. Now as $\partial_c(\xi) = \nabla\zeta + \theta_c \circ \xi$:

$$\|\nabla\xi_i\|_0 \leq \|\partial_c(\xi_i)\|_0 + \|\theta_c \circ \xi_i\|_0 ,$$

$$\kappa\|\partial_c(\xi_i)\|_0^2 \leq <(G_c \circ \xi_i)\cdot\partial_c(\xi_i), \partial_c(\xi_i)>_0 \leq 2a ,$$

for some $\kappa > 0$ so $\|\xi_i\|_1$ is bounded. We will prove convergence of this sequence in $H^1(c^*\tau)$ using the following estimates:

$$\|\nabla\xi - \nabla\eta\|_0 \leq \|\partial_c(\xi) - \partial_c(\eta)\|_0 + \|\theta_c \circ \xi - \theta_c \circ \eta\|_0 ,$$

$$\begin{aligned} 2\kappa\|\partial_c(\xi) - \partial_c(\eta)\|_0^2 &\leq <(G_c \circ \xi)\cdot(\partial_c(\xi) - \partial_c(\eta)), \partial_c(\xi) - \partial_c(\eta)>_0 + \\ &\quad + <(G_c \circ \eta)\cdot(\partial_c(\xi) - \partial_c(\eta)), \partial_c(\xi) - \partial_c(\eta)>_0 \\ &= 2dE_c(\xi)\cdot(\xi - \eta) - 2dE_c(\eta)\cdot(\xi - \eta) + \\ &\quad - <(G_c \circ \xi - G_c \circ \eta)\cdot(\partial_c(\xi) + \partial_c(\eta)), \partial_c(\xi) - \partial_c(\eta)>_0 + \\ &\quad + 2<(G_c \circ \xi)\cdot\partial_c(\xi), \theta_c \circ \xi - \theta_c \circ \eta - (D_2\theta_c \circ \xi)\cdot(\xi - \eta)>_0 + \\ &\quad - 2<(G_c \circ \eta)\cdot\partial_c(\eta), \theta_c \circ \xi - \theta_c \circ \eta - (D_2\theta_c \circ \eta)\cdot(\xi - \eta)>_0 + \\ &\quad - \tfrac{1}{2}<(D_2G_c \circ \xi)\cdot(\xi - \eta, \partial_c(\xi)), \partial_c(\xi)>_0 \\ &\quad + \tfrac{1}{2}<(D_2G_c \circ \xi)\cdot(\xi - \eta, \partial_c(\eta)), \partial_c(\eta)>_0 . \end{aligned}$$

Now as ξ_i is bounded in H^1 and converging in C^0 and $dE_c(\xi_i) \to 0$ we obtain $\|\nabla\xi_i - \nabla\xi_j\| \to 0$ for $i, j \to \infty$ by replacing ξ, η by ξ_i, ξ_j. So ξ_i is a Cauchy sequence in $H^1(c^*\tau M)$ and it follows that ξ_0 is the limit and $dE_c(\xi_0) = 0$.

The formula for the Hessian in Theorem 4 is most easily obtained by taking the variational derivative of E in the directions given by ξ, η (for justification note the remark in §2(b)) and the rest of §1 does not need further explanation.

REFERENCES

[1] R. Bott, "Non-degenerate critical manifolds," *Ann. of Math. (2) 60* (1954), 248-261.

[2] R. Bott and H. Samelson, "Applications of the theory of Morse to symmetric spaces," *Am. J. of Math.*, Vol. *80* (1958), 964-1029.

[3] J. Eells, Jr., "On the geometry of function spaces," *Symp. Inter. de Topologia Alg. Mexico* (1956); 1958, 303-308.

[4] ———, "Alexander-Pontrjagin duality in function spaces," *Proc. of Symp. in Pure Math.*, Vol. 888, A. M. S. (1961).

[5] H. I. Eliasson, "On the geometry of manifolds of maps," *J. Diff. Geom.* 2 (1967).

[6] A. I. Fet, "Variational problems on closed manifolds," *Amer. Math. Soc. Transl.* no. 90 (1953).

[7] R. S. Palais, "Lectures on the differential topology of infinite-dimensional manifolds," Mim. notes at Brandeis by S. Greenfield (1964-65).

[8] ———, "Morse theory on Hilbert manifolds," *Topology 2* (1963), 299-340.

[9] R. S. Palais and S. Smale, "A generalized Morse theory," *Bull. Amer. Math. Soc.*, *70* (1964), 165-172.

[10] W. Meyer, "Kritische Mannigfaltigkeiten in Hilbertmannigfaltigkeiten," *Math. Annalen*, *170* (1967), 45-66.

[11] A. S. Švarc, "Homology of spaces of closed curves," *Dokl. Akad. Nauk. SSSR.*, Vol. *117* (1957), 769-772.

[12] A. Wasserman, "Morse theory for G-manifolds," *Bull. Amer. Math. Soc.*, *71* (1965), 384-388.

[13] N. Steenrod, *The Topology of Fibre Bundles*, Princeton Univ. Press, Princeton, N. J., 1951.

BROWN UNIVERSITY

COVERING PROPERTIES OF CONVEX SETS AND FIXED POINT THEOREMS IN TOPOLOGICAL VECTOR SPACES*

by

KY FAN

§1. *Introduction.* In the finite dimensional case, it is well-known that the classical results of Sperner [10] and Knaster-Kuratowski-Mazurkiewicz [9] on covering properties of n-simplexes are closely related to the Brouwer fixed point theorem. Related results on covering properties of convex sets in general topological vector spaces have been given by Berge [2] and Ghouila-Houri [6]. In §2 of the present paper, we shall apply the classical results of Sperner and Knaster-Kuratowski-Mazurkiewicz to prove three theorems on common fixed points or coincidences for a family of continuous mappings on an n-simplex. A special case of Theorems 1 and 2 will be used in §3 to obtain new results on covering properties of convex sets in topological vector spaces, not necessarily of finite dimension. Theorem 4 sharpens a theorem of Ghouila-Houri. The covering theorems of Sperner and Knaster-Kuratowski-Mazurkiewicz are extended and sharpened to Theorems 6 and 5 respectively.

In the literature, there are fixed point theorems for upper semi-continuous set-valued mappings, In §4 we consider lower semi-continuous set-valued mappings on a convex set, not necessarily compact, in a locally convex topological vector space, and prove a theorem on almost fixed points for such mappings by using the covering theorem of Knaster-Kuratoswki-

* Work supported in part by the National Science Foundation, Grant GP-5578.

Mazurkiewicz. In §5 we again use this basic theorem to give a short proof of a generalized form of Tychonoff's fixed point theorem [11].

Throughout the paper all topological vector spaces are assumed to be separated (i.e., Hausdorff) topological vector spaces over the real or complex field.

§ 2. *Common fixed points for a family of continuous mappings of an n-simplex. In the following theorem concerning common fixed points for a family of continuous mappings* $\{g_\nu\}_{\nu \in N}$ *of an n-simplex* S, *we consider, instead of the identity mapping of* S, *a more general mapping* f. *This generality will be needed in* §3.

THEOREM 1. *Let* $T = v_0 v_1 \cdots v_p$ *be a* p*-simplex, and let* S *be the* n*-face* $v_0 v_1 \cdots v_n$ *of* T, *where* $0 \leq n \leq p$. *Let* $f\colon S \to T$ *be a continuous mapping:*

$$(1) \qquad f(x) = \sum_{j=0}^{p} a_j(x) v_j \text{ with } a_j(x) \geq 0, \quad \sum_{j=0}^{p} a_j(x) = 1, \text{ for } x \in S$$

such that the following conditions (2), (3), (4) *are fulfilled:*

(2) $f(S)$ *is contained in the* n*-skeleton of* T *(i.e., for each* $x \in S$, *at most* $n+1$ *of* $a_j(x)$, $0 \leq j \leq p$, *are positive).*

(3) $f(S)$ *is disjoint from the face* $v_{n+1} v_{n+2} \cdots v_p$ *of* T *(i.e.,* $\Sigma_{i=0}^{n} a_i(x) > 0$ *for every* $x \in S$*).*

(4) $a_i(x) = 0$ *for each* $i = 0, 1, \ldots, n$ *and for any point* x *in the* $(n-1)$*-face* $v_0 \cdots v_{i-1} v_{i+1} \cdots v_n$ *of* S.

Let $\{g_\nu\}_{\nu \in N}$ *be an indexed family of continuous mappings from* S *into* S:

$$(5) \quad g_\nu(x) = \sum_{i=0}^{n} \beta_{\nu,i}(x) v_i \text{ with } \beta_{\nu,i}(x) \geq 0, \quad \sum_{i=0}^{n} \beta_{\nu,i}(x) = 1, \text{ for } x \in S,$$

satisfying the following condition:

(6) *For every* $x \in S$, *there is an index* $i \in \{0, 1, \ldots, n\}$, *depending on* x, *such that* $\alpha_i(x) \leq \beta_{\nu,i}(x)$ for all $\nu \in N$.

Then there exists a point $\hat{x} \in S$ *such that*

(7) $$f(\hat{x}) = g_\nu(\hat{x}) \text{ for all } \nu \in N .$$

Proof: For $0 \leq i \leq n$, let A_i denote the set of all $x \in S$ satisfying the following condition:

(8) $$\alpha_i(x) \left[1 + \sum_{j=n+1}^{p} \alpha_j(x)\right] \leq \beta_{\nu,i}(x) \left[1 - \sum_{j=n+1}^{p} \alpha_j(x)\right] \text{ for all } \nu \in N .$$

By property (4) of f, we have

(9) $$v_0 \cdots v_{i-1} v_{i+1} \cdots v_n \subset A_i \quad \text{for } 0 \leq i \leq n .$$

For each $x \in S$, $f(x)$ is in the n-skeleton of T, so at least $p-n$ of $\alpha_j(x)$ $(0 \leq j \leq p)$ are 0. If $\alpha_{n+1}(x), \alpha_{n+2}(x), \ldots, \alpha_p(x)$ are not all 0, then $\alpha_i(x) = 0$ for at least one index $i \in \{0, 1, \ldots, n\}$, and therefore $x \in \cup_{i=0}^{n} A_i$. On the other hand, if $\alpha_{n+1}(x) = \alpha_{n+2}(x) = \cdots = \alpha_p(x) = 0$ or if $n = p$, then condition (6) implies that (8) is satisfied for at least one index $i \in \{0, 1, \ldots, n\}$, and therefore again $x \in \cup_{i=0}^{n} A_i$. This shows that

(10) $$\bigcup_{i=0}^{n} A_i = S .$$

We claim that

(11) $$\bigcap_{i=0}^{n} A_i \neq \emptyset .$$

Suppose that the contrary is true, i.e., $\cup_{i=0}^{n} A_i' = S$, where A_i' denotes the complement of A_i in S. As A_i' are open in S, we can find closed sets B_i $(0 \leq i \leq n)$ such that $B_i \subset A_i'$ $(0 \leq i \leq n)$ and $\cup_{i=0}^{n} B_i = S$. Then (9) implies

(12) $$v_0 \cdots v_{i-1} v_{i+1} \cdots v_n \cap B_i = \emptyset \text{ for } 0 \leq i \leq n .$$

By a classical result of Sperner [10], the $n+1$ closed sets B_i $(0 \leq i \leq n)$ covering S and satisfying (12) must have a non-empty intersection. Then we would have $\cap_{i=0}^{n} A_i' \supset \cap_{i=0}^{n} B_i \neq \emptyset$ and $\cup_{i=0}^{n} A_i \neq S$, against (10). This proves (11).

Consider now a point $\hat{x} \in \cap_{i=0}^{n} A_i$. Summing (8) over $i = 0, 1, \ldots, n$ for this $\hat{x}$, we obtain

$$(13) \qquad \left[1 - \sum_{j=n+1}^{p} \alpha_j(\hat{x})\right] \cdot \left[1 + \sum_{j=n+1}^{p} \alpha_j(\hat{x})\right] \leq 1 - \sum_{j=n+1}^{p} \alpha_j(\hat{x}) .$$

By hypothesis (3), we have $\Sigma_{j=n+1}^{p} \alpha_j(\hat{x}) < 1$. Therefore (13) implies $\alpha_j(\hat{x}) = 0$ for $n+1 \leq j \leq p$. Then (8) for $x = \hat{x}$ becomes

$$\alpha_i(\hat{x}) \leq \beta_{\nu,i}(\hat{x}) \quad \text{for } \nu \in N \text{ and } 0 \leq i \leq n .$$

Combining this with $\Sigma_{i=0}^{n} \alpha_i(\hat{x}) = \Sigma_{i=0}^{n} \beta_{\nu,i}(\hat{x}) = 1$, we get

$$\alpha_i(\hat{x}) = \beta_{\nu,i}(\hat{x}) \quad \text{for } \nu \in N \text{ and } 0 \leq i \leq n ,$$

which is (7). Theorem 1 is thus proved.

Conditions (2), (3), (4) concerning f are clearly satisfied by the identity mapping of S. Observe also that, because $\Sigma_{i=0}^{n} \alpha_i(x) \leq 1 = \Sigma_{i=0}^{n} \beta_{\nu,i}(x)$, condition (6) will be automatically satisfied, if the family $\{g_\nu\}_{\nu \in N}$ consists of a single mapping. In this case, we have the following result which will be used in §3.

COROLLARY 1. *Let* $T = v_0 v_1 \cdots v_p$ *be a* p-*simplex and let* S *be the* n-*face* $v_0 v_1 \cdots v_n$ *of* T, *where* $0 \leq n \leq p$. *Let* $f\colon S \to T$ *be a continuous mapping satisfying conditions* (2), (3) *and* (4). *Then for every continuous mapping* $g\colon S \to S$, *there exists a point* $x \in S$ *such that* $f(x) = g(x)$. *In particular:* $S \subset f(S)$, *and* f *has a fixed point.*

Proof: That $S \subset f(S)$ is seen by considering those g which map S to a single point of S. On the other hand, if we take the identity mapping of S as g, then the existence of a fixed point of f follows.

The next result is obtained by replacing condition (4) of Theorem 1 by (14) below.

THEOREM 2. *Let* $T = v_0 v_1 \cdots v_p$ *be a* p-*simplex, and let* S *be the* n-*face* $v_0 v_1 \cdots v_n$ *of* T, *where* $0 \le n \le p$. *Let* $f\colon S \to T$ *be a continuous mapping represented by* (1) *and satisfying conditions* (2), (3) *and*

(14) *for any* $k+1$ *indices* $0 \le i_0 < i_1 < \cdots < i_k \le n$, *where* $0 \le k \le n-1$, *and for any point* x *in the* k-*face* $v_{i_0} v_{i_1} \cdots v_{i_k}$ *of* S, *at least one of* $\alpha_{i_0}(x), \alpha_{i_1}(x), \ldots, \alpha_{i_k}(x)$ *vanishes.*

Let $\{g_\nu\}_{\nu \in N}$ *be a family of continuous mappings from* S *into* S, *represented by* (5) *and satisfying condition* (6). *Then there exists a point* $\hat{x} \in S$ *satisfying* (7).

Proof: As in the proof of Theorem 1, we consider the set A_i $(0 \le i \le n)$ formed by all $x \in S$ satisfying (8). By property (14) of f, we have

$$v_{i_0} v_{i_1} \cdots v_{i_k} \subset A_{i_0} \cup A_{i_1} \cup \cdots \cup A_{i_k} \tag{15}$$

for

$$0 \le k \le n-1 \text{ and } 0 \le i_0 \le i_1 < \cdots < i_k \le n .$$

As in the proof of Theorem 1, we have (10) as a consequence of (2) and (6). Now by a classical theorem of Knaster-Kuratowski-Marzurkiewicz [9], the $n+1$ closed subsets A_i $(0 \le i \le n)$ of the n-simplex S satisfying (15) and (10) must have a non-empty intersection. Then, by the same argument used in the proof of Theorem 1, we see that every point $\hat{x}$ in the intersection $\bigcap_{i=0}^{n} A_i$ satisfies (7).

COROLLARY 2. *Let* $T = v_0 v_1 \cdots v_p$ *be a* p-*simplex, and let* S *be the* n-*face* $v_0 v_1 \cdots v_n$ *of* T, *where* $0 \le n \le p$. *Let* $f\colon S \to T$ *be a continuous mapping satisfying conditions* (2), (3), *and* (14). *Then for every continuous mapping* $g\colon S \to S$, *there exists a point* $x \in S$ *such that* $f(x) = g(x)$. *In particular:* $S \subset f(S)$, *and* f *has a fixed point.*

The following variant of Theorem 2 is obtained by replacing condition (14) by (16) below.

THEOREM 3. *Let* $T = v_0 v_1 \cdots v_p$ *be a* p-*simplex, and let* S *be the* n-*face* $v_0 v_1 \cdots v_n$ *of* T, *where* $0 \leq n \leq p$. *Let* $f: S \to T$ *be a continuous mapping represented by* (1) *and satisfying conditions* (2) *and* (3). *Let* $\{g_\nu\}_{\nu \in N}$ *be a family of continuous mappings from* S *into* S, *represented by* (5) *and satisfying* (6) *and the following condition:*

(16) *For each* $i = 0, 1, \ldots, n$ *and for every point* x *in the* $(n-1)$-*face* $v_0 \cdots v_{i-1} v_{i+1} \cdots v_n$ *of* S, *there is a* $\nu \in N$ *such that* $\alpha_i(x) > \beta_{\nu,i}(x)$.

Then there exists a point $\hat{x} \in S$ *satisfying* (7).

Proof: Let A_i be again the set of all $x \in S$ satisfying (8). As in the proof of Theorem 2, it suffices to show that $\bigcap_{i=0}^n A_i \neq \emptyset$. Condition (16) implies that for each $i = 0, 1, \ldots, n$, A_i is disjoint from the $(n-1)$-face $v_0 \cdots v_{i-1} v_{i+1} \cdots v_n$ of S. On the other hand, (10) holds as a consequence of (2) and (6). Thus the desired relation $\bigcap_{i=0}^n A_i \neq \emptyset$ follows from Sperner's result [10] used in the proof of Theorem 1.

We have already observed that condition (6) in Theorems 1, 2, 3 is automatically satisfied if the family $\{g_\nu\}_{\nu \in N}$ consists of a single mapping. Notice also that conditions (2) and (3) will be automatically satisfied if f maps S into S.

§3. *Covering properties of convex sets.* In this section, we prove some new covering properties of convex sets in topological vector spaces, which need not be finite dimensional.

THEOREM 4. *Let* C *be a convex set in a topological vector space. Let* $A_0, A_1, \ldots, A_n$ *and* B *be subsets of* C *satisfying the following three conditions:*

(17) $$B \cup \bigcup_{i=0}^{n} A_i = C,$$

(18) $$\bigcap_{i=0}^{n} A_i = \emptyset ,$$

(19) *Each of* $A_0, A_1, \ldots, A_n$ *and* B *is closed in* C.

If there exist convex subsets $C_0, C_1, \ldots, C_n$ *of* C *such that*

(20) $$C_i \subset A_i \quad \text{for } 0 \leq i \leq n ,$$

(21) *the intersection of every* n *of the* $n+1$ *sets* $C_0, C_1, \ldots, C_n$ *is non-empty; then*

(22) $$B \cap \left(\bigcap_{i \in I'} A_i \right) \cap \left(\bigcap_{i \in I} A_i{}' \right) \neq \emptyset$$

for every non-empty subset I *of* $\{0, 1, \ldots, n\}$. *Here* I' *denotes the complement of* I *in* $\{0, 1, \ldots, n\}$, *and* $A_i{}'$ *denotes the complement of* A_i *in* C.

Proof: By (21), there exist $n+1$ points $w_0, w_1, \ldots, w_n$ such that $w_j \in C_i$ whenever $i \neq j$. Consider an n-simplex $S = v_0 v_1 \cdots v_n$ and the continuous mapping $\phi: S \to C$ defined by

(23) $$\phi\left(\sum_{i=0}^{n} a_i v_i \right) = \sum_{i=0}^{n} a_i w_i \quad \text{for } a_i \geq 0, \ \sum_{i=0}^{n} a_i = 1 .$$

Let $D_i = \phi^{-1}(A_i)$ $(0 \leq i \leq n)$, and $D_{n+1} = \phi^{-1}(B)$. By (19), D_i $(0 \leq i \leq n+1)$ are closed subsets of S. For $x \in S$, let $\delta_i(x)$ denote the distance from x to D_i $(0 \leq i \leq n+1)$. Let $T = v_0 v_1 \cdots v_n v_{n+1}$ be an $(n+1)$-simplex containing $S = v_0 v_1 \cdots v_n$ as an n-face. Define $f: S \to T$ by

(24) $$f(x) = \left[\sum_{j=0}^{n+1} \delta_j(x) \right]^{-1} \sum_{i=0}^{n+1} \delta_i(x) v_i \quad \text{for } x \in S .$$

As (18) implies $\bigcap_{i=0}^{n} D_i = \emptyset$, we have $\sum_{i=0}^{n} \delta_i(x) > 0$ for every $x \in S$. Thus f is well-defined and the vertex v_{n+1} is not in $f(S)$. By (17) we have $\bigcup_{i=0}^{n+1} D_i = S$. Therefore, for every $x \in S$, at least one of $\delta_i(x)$ $(0 \leq i \leq n+1)$ is 0. This means that f maps S into the n-skeleton of T. Thus f satisfies conditions (2) and (3).

For each $i = 0, 1, \ldots, n$, the convex hull of $w_0, \ldots, w_{i-1}, w_{i+1}, \ldots, w_n$ is contained in C_i. If x is a point in the $(n-1)$-face $v_0 \cdots v_{i-1} v_{i+1} \cdots v_n$, of S, then $\phi(x)$ is in the convex hull of $w_0, \ldots, w_{i-1}, w_{i+1}, \ldots, w_n$, so $\phi(x) \in C_i \subset A_i$. Therefore for any $i = 0, 1, \ldots, n$ and for every $x \in v_0 \cdots v_{i-1} v_{i+1} \cdots v_n$, we have $x \in D_i$ and $\delta_i(x) = 0$. Thus f satisfies condition (4).

By Corollary 1, we have $S \subset f(S)$. In other words, for any given non-negative numbers $\gamma_i \geq 0$ $(0 \leq i \leq n)$ with $\Sigma_{i=0}^n \gamma_i = 1$, there is a point $x \in S$ for which

$$\delta_i(x) = \gamma_i \sum_{j=0}^{n} \delta_j(x) \text{ for } 0 \leq i \leq n, \quad \delta_{n+1}(x) = 0 . \tag{25}$$

In particular, for any non-empty subset I of $\{0, 1, \ldots, n\}$ if the γ's are so chosen that $\gamma_i > 0$ for $i \in I$, and $\gamma_i = 0$ for $i \in I'$, then every point $x \in S$ satisfying (25) is in the intersection

$$D_{n+1} \cap \left(\bigcap_{i \in I'} D_i \right) \cap \left(\bigcap_{i \in I} D_i' \right),$$

where D_i' denotes the complement of D_i in S. Then it follows that (22) holds for every non-empty subset I of $\{0, 1, \ldots, n\}$. This completes the proof.

COROLLARY 3. *In a topological vector space* E, *let* $C_0, C_1, \ldots, C_n$ *be* $n+1$ *closed convex sets such that* $\bigcap_{i=0}^n C_i = \emptyset$ *and the intersection of every* n *of them is non-empty. If* B *is a closed set in* E *such that* $B \cup \bigcup_{i=0}^n C_i$ *is convex, then*

$$B \cap \left(\bigcap_{i \in I'} C_i \right) \cap \left(\bigcap_{i \in I} C_i' \right) \neq \emptyset \tag{26}$$

for every non-empty subset I *of* $\{0, 1, \ldots, n\}$. *Here* C_i' *denotes the complement of* C_i *in* E.

Clearly Corollary 3 is a special case of Theorem 4. Corollary 3 sharpens a theorem of Ghouila-Houri [6], who proved under the same hypothesis that the intersection of any n of the sets $C_0, C_1, \ldots, C_n$ contains a point of B. For some interesting consequences of Ghouila-Houri's theorem, see [2], [3, pp. 123-124], [6].

THEOREM 5. *Let* C *be a convex set in a topological vector space. Let* $A_0, A_1, \ldots, A_n$ *and* B *be subsets of* C *satisfying conditions* (17), (18) *and* (19). *If there are* $n+1$ *points* $w_0, w_1, \ldots, w_n$ *in* C *such that:*

(27) *for any* $k+1$ *indices* $0 \leq i_0 < i_1 < \cdots < i_k \leq n$, *where* $0 \leq k \leq n-1$, *the convex hull of* $w_{i_0}, w_{i_1}, \ldots, w_{i_k}$ *is contained in the union*

$$A_{i_0} \cup A_{i_1} \cup \cdots \cup A_{i_k},$$

then (22) *holds for every non-empty subset* I *of* $\{0, 1, \ldots, n\}$.

Proof: Consider an $(n+1)$-simplex $T = v_0 v_1 \cdots v_n v_{n+1}$ and its n-face $S = v_0 v_1 \cdots v_n$. Using the $n+1$ points $w_0, w_1, \ldots, w_n$ having property (27), we define $\phi: S \to C$ by (23). Let $D_i = \phi^{-1}(A_i)$ $(0 \leq i \leq n)$, and $D_{n+1} = \phi^{-1}(B)$. For $x \in S$, let $\delta_i(x)$ denote the distance from x to D_i $(0 \leq i \leq n+1)$. Next define $f: S \to T$ by (24). Then as in the proof of Theorem 4, f satisfies conditions (2) and (3).

Consider now any $k+1$ indices $0 \leq i_0 < i_1 < \cdots < i_k \leq n$, where $0 \leq k \leq n-1$, and a point $x \in v_{i_0} v_{i_1} \cdots v_{i_k}$. By (27), $\phi(x)$ is the union $A_{i_0} \cup A_{i_1} \cup \cdots \cup A_{i_k}$, so x is in $D_{i_0} \cup D_{i_1} \cup \cdots \cup D_{i_k}$. Therefore one of $\delta_{i_0}(x), \delta_{i_1}(x), \ldots, \delta_{i_k}(x)$ is 0. This shows that f has property (14).

By Corollary 2, we have $S \subset f(S)$. Then the proof can be completed as in the proof of Theorem 4.

Theorem 5 remains valid if instead of being convex, C is only assumed to contain the convex hull of $w_0, w_1, \ldots, w_n$. In case $B = \emptyset$, Theorem 5 becomes a generalized form of Knaster-Kuratowski-Mazurkiewicz's theorem. The basic result is stated here as

COROLLARY 4. *In a topological vector space* E, *let* $A_0, A_1, \ldots, A_n$ *be* $n+1$ *sets, each of which is closed in the union* $\bigcup_{i=0}^{n} A_i$. *Let* $w_0, w_1, \ldots, w_n$ *be* $n+1$ *points in* E. *If, for every non-empty subset* I *of* $\{0, 1, \ldots, n\}$, *the convex hull of the set* $\{w_i : i \in I\}$ *is contained in* $\bigcup_{i \in I} A_i$, *then* $\bigcap_{i=0}^{n} A_i \neq \emptyset$.

The next corollary, which was already given in [4], follows directly from Corollary 4.

COROLLARY 5. *Let* X *be an arbitrary set in a topological vector space* E. *To each* $x \in X$, *let a closed set* $A(x)$ *in* E *be given such that* $A(x)$ *is compact for at least one* $x \in X$. *If the convex hull of any finite subset* $\{x_1, x_2, \ldots, x_n\}$ *of* X *is contained in* $\bigcup_{i=1}^{n} A(x_i)$, *then* $\bigcap_{x \in X} A(x) \neq \emptyset$.

The following result extends and sharpens the covering theorem of Sperner.

THEOREM 6. *Let* C *be a convex set in a topological vector space. Let* $A_0, A_1, \ldots, A_n$ *and* B *be subsets of* C *satisfying conditions* (17), (18) *and* (19). *If there exist* $n+1$ *convex subsets* $C_0, C_1, \ldots, C_n$ *of* C *satisfying* (21) *and*

$$C_i \cap (B \cup A_i) = \emptyset \quad \text{for} \quad 0 \leq i \leq n, \tag{28}$$

then (22) *holds for every non-empty subset* I *of* $\{0, 1, \ldots, n\}$.

Proof: By (21), there are $n+1$ points $w_0, w_1, \ldots, w_n$ such that $w_i \in C_j$ whenever $i \neq j$. Consider any $k+1$ indices $0 \leq i_0 < i_1 < \cdots < i_k \leq n$, where $0 \leq k \leq n-1$. Let $\{i_{k+1}, i_{k+2}, \ldots, i_n\}$ be the complement of $\{i_0, i_1, \ldots, i_k\}$ in the set $\{0, 1, \ldots, n\}$. Then the convex hull of $w_{i_0}, w_{i_1}, \ldots, w_{i_k}$ is contained in the intersection $C_{i_{k+1}} \cap C_{i_{k+2}} \cap \cdots \cap C_{i_n}$. In view of (17) and (28), this intersection is contained in the union $A_{i_0} \cup A_{i_1} \cup \cdots \cup A_{i_k}$. Hence the points $w_0, w_1, \ldots, w_n$ have property (27), and the result follows from Theorem 5.

§4. *Almost fixed points for lower semi-continuous set-valued mappings.* Let X, Y be two topological spaces, and let $\mathcal{P}(Y)$ denote the collection of all subsets of Y. A set-valued mapping $f: X \to \mathcal{P}(Y)$ is said to be *lower semi-continuous* on X, if for every point $x_0 \in X$ and any open set G in Y such that $f(x_0) \cap G \neq \emptyset$. there is a neighborhood U of x_0 in X such that $f(x) \cap G \neq \emptyset$ for every $x \in U$. In other words, $f: X \to \mathcal{P}(Y)$ is lower semi-continuous on X, if for every open set G in Y, the set $\{x \in X: f(x) \cap G \neq \emptyset\}$ is open in X (see [1, p. 114]).

THEOREM 7. *Let X be a non-empty convex set in a locally convex topological vector space E. Let $f: X \to \mathcal{P}(E)$ be a lower semi-continuous mapping such that $f(x)$ is a non-empty convex subset of E for each $x \in X$. If there is a precompact subset Y of X such that $f(x) \cap Y \neq \emptyset$ for each $x \in X$, then for every neighborhood W of 0 in E, there exists a point $\hat{x} \in X$ such that $f(\hat{x}) \cap (\hat{x} + W) \neq \emptyset$.*

Notice that X need not be compact, nor closed.

Proof: Because E is locally convex, it suffices to consider those neighborhoods of 0 in E which are open and convex. Consider an arbitrarily fixed open convex neighborhood W of 0 in E. For each $x \in X$, let A(x) be the set of all $y \in X$ such that $f(y) \cap (x+W) = \emptyset$. Since $x+W$ is open, A(x) is closed in X for each $x \in X$, by the lower semi-continuity of f. For the precompact subset Y of X, we can find a finite set $\{x_1, x_2, \ldots, x_n\} \subset X$ such that $Y \subset \bigcup_{i=1}^{n} (x_i + W)$. Then for every $x \in X$, we have $f(x) \cap \bigcup_{i=1}^{n} (x_i + W) \neq \emptyset$. Hence $\bigcap_{i=1}^{n} A(x_i) = \emptyset$. As $A(x_i)$ is closed in X, it is closed in $\bigcup_{i=1}^{n} A(x_i)$.

We now apply Corollary 4. There is a non-empty subset I of $\{1, 2, \ldots, n\}$ such that the convex hull of $\{x_i: i \in I\}$ is not contained in $\bigcup_{i \in I} A(x_i)$. For notational simplicity, we may assume that $I = \{1, 2, \ldots, p\}$ for some $p \leq n$. Let $\alpha_i \geq 0$ $(1 \leq i \leq p)$ be such that $\sum_{i=1}^{p} \alpha_i = 1$ and $\sum_{i=1}^{p} \alpha_i x_i \notin \bigcup_{i=1}^{p} A(x_i)$. Let $\hat{x} = \sum_{i=1}^{p} \alpha_i x_i$. Then $f(\hat{x}) \cap (x_i + W) \neq \emptyset$ for $1 \leq i \leq p$, or what is the same, $x_i \in f(\hat{x}) - W$ for $1 \leq i \leq p$. As $f(\hat{x}) - W$ is convex, it follows that $\hat{x} \in f(\hat{x}) - W$, i.e., $f(\hat{x}) \cap (\hat{x} + W) \neq \emptyset$; and the proof is complete.

In case $f(x)$ is a single point for each $x \in X$, Theorem 7 becomes the following:

COROLLARY 6. *Let X be a non-empty convex set in a locally convex topological vector space E. Let $f\colon X \to X$ be a continuous mapping such that $f(X)$ is precompact. Then for every neighborhood W of 0 in E, there exists an $\hat{x} \in X$ such that $\hat{x} - f(\hat{x}) \in W$.*

Corollary 6 may be regarded as a generalization of Tychonoff's fixed point theorem [11] to non-compact (or precompact) convex sets. In another connection, Corollary 6 should be compared with a result of Klee (Theorem 3 in [8]) concerning ε-continuous mappings of a compact convex set in a normed vector space.

§5 §5. *A short proof of a generalized Tychonoff's fixed point theorem.*

In this section we show that Knaster-Kuratowski-Mazurkiewicz's theorem (stated above as Corollary 4) implies in a very direct and simple way the fixed point theorem of Tychonoff. In fact we shall give a short proof of the following more general result, which was obtained in [5] as an extension of a theorem of Iohvidov [7].

THEOREM 8. *In a locally convex topological vector space E, let X be a non-empty compact convex set and K a non-empty closed convex set. Let $g\colon X \times X \to E$ be a continuous mapping such that for each $x \in X$ and any convex set C in E, the set $\{y \in X\colon g(x, y) \in C\}$ is convex. If for every $x \in X$, there is a $y \in X$ with $g(x, y) \in K$, then there exists an $\hat{x} \in X$ such that $g(\hat{x}, \hat{x}) \in K$.*

In case $K = \{0\}$ and $g\colon X \times X \to E$ is defined by $g(x, y) = f(x) - y$ with a continuous mapping $f\colon X \to X$, Theorem 8 reduces to the classical theorem of Tychonoff.

Proof: Let $\mathcal{V}$ be the system of all open convex neighborhoods of 0 in E. For each $V \in \mathcal{V}$, let F_V denote the closed set formed by all $(x, y) \in X \times X$ satisfying $x - y \in \overline{V}$ and $g(x, y) \in \overline{K + V}$. If we can prove that $F_V \neq \emptyset$ for each $V \in \mathcal{V}$, then it will follow that

$$\bigcap_{i=1}^{n} F_{V_i} \supset F_{V_1 \cap V_2 \cap \cdots \cap V_n} \neq \emptyset$$

for any finite number of $V_i \in \mathcal{V}$. The compactness of $X \times X$ will then imply $\bigcap_{V \in \mathcal{V}} F_V \neq \emptyset$. Clearly any point $(\hat{x}, \hat{y}) \in \bigcap_{V \in \mathcal{V}} F_V$ will satisfy $\hat{x} = \hat{y}$ and $g(\hat{x}, \hat{x}) \in K$.

From now on, we consider an arbitrarily fixed $v \in \mathcal{V}$. For each $(x, y) \in X \times X$, let $A(x, y)$ be the set of all $(u, v) \in X \times X$ such that either $g(u, y) \notin K + V$ or $v \notin x - V$. Since $K + V$ and $x - V$ are both open in E, the continuity of g implies that $A(x, y)$ is a closed subset of $X \times X$.
For any given $(u, v) \in X \times X$, we can take $x = v$, and by hypothesis we can choose $y \in X$ such that $g(u, y) \in K$. For such a choice of (x, y), we have $g(u, y) \in K + V$ and $v \in x - V$, so $(u, v) \notin A(x, y)$. This shows that $\bigcap_{(x,y) \in X \times X} A(x, y) = \emptyset$.

Applying Corollary 5, we can find a finite number of points $(x_j, y_j) \in X \times X$ $(1 \leq j \leq p)$ and $\alpha_j \geq 0$ $(1 \leq j \leq p)$ with $\Sigma_{j=1}^{p} \alpha_j = 1$ such that

$$\sum_{j=1}^{p} \alpha_j (x_j, y_j) \notin \bigcup_{j=1}^{p} A(x_j, y_j) .$$

Let $x_0 = \Sigma_{j=1}^{p} \alpha_j x_j$, $y_0 = \Sigma_{j=1}^{p} \alpha_j y_j$. Then

(29) $$g(x_0, y_j) \in K + V \qquad (1 \leq j \leq p),$$

(30) $$y_0 \in x_j - V \qquad (1 \leq j \leq p) .$$

Since $K + V$ is convex, the set $\{y \in X: g(x_0, y) \in K + V\}$ is convex by hypothesis. Therefore (29) implies $g(x_0, y_0) \in K + V$. On the other hand, (30) means that $x_1, x_2, \ldots, x_p$ are in the convex set $y_0 + V$. Therefore $x_0 \in y_0 + V$. We have thus $(x_0, y_0) \in F_V$. This shows that $F_V \neq \emptyset$ and the proof is complete.

REFERENCES

[1] C. Berge, *Espaces topologiques, Fonctions multivoques*, Dunod, Paris, 1959.

[2] _____, "Sur une propriété combinatoire des ensembles convexes," *C. R. Acad. Sci., Paris, 248* (1959), 2698-2699.

[3] L. Danzer, B. Grünbaum and V. Klee, "Helly's theorem and its relatives," Convexity, *Proc. Sympos. Pure Math., 7,* edited by V. Klee, Amer. Math. Soc., Providence, R. I., 1963, 101-177.

[4] K. Fan, "A generalization of Tychonoff's fixed point theorem," *Math. Annalen, 142* (1961), 305-310.

[5] _____, "Applications of a theorem concerning sets with convex sections," *Math. Annalen, 163* (1966), 189-203.

[6] A. Ghouila-Houri, "Sur l'étude combinatoire des familles de convexes," *C. R. Acad. Sci.*, Paris, *252* (1961), 494-496.

[7] I. S. Iohvidov, "On a lemma of Ky Fan generalizing the fixed point principle of A. N. Tihonov" (Russian), *Dokl. Akad. Nauk SSSR, 159* (1964), 501-504. English transl.: *Soviet Math., 5* (1964), 1523-1526.

[8] V. Klee, "Stability of the fixed-point property," *Colloquium Math., 8* (1961), 43-46.

[9] B. Knaster, C. Kuratowski and S. Mazurkiewicz, "Ein Beweis des Fixpunktsatzes für n-dimensionale Simplexe," *Fund. Math., 14* (1929), 132-137.

[10] E. Sperner, "Neuer Beweis für die Invarianz der Dimensionszahl und des Gebietes," *Abhandl. Math. Seminar Univ. Hamburg, 6* (1928), 265-272.

[11] A. Tychonoff, "Ein Fixpunktsatz," *Math. Annalen, 111* (1935), 767-776.

UNIVERSITY OF CALIFORNIA
SANTA BARBARA, CALIFORNIA

ON THE COHOMOLOGY THEORY IN LINEAR NORMED SPACES *

KAZIMIERZ GEBA AND ANDRZEJ GRANAS

Let E be a linear normed space and $(\mathfrak{L}^{\sim}, \sim)$ be the h-category whose objects are closed bounded subsets of E, whose morphisms are compact vector fields and in which the relation $\sim$ means the compact homotopy of fields. This category, introduced in 1933 by Leray and Schauder [4] in their study of topological degree plays a basic role in topology of an infinite dimensional normed space E.

The main theorem which we intend to present here is the following: for each $n \geq 1$ and a coefficient group G, there exists an h-functor $H^{\infty - n}$ from the category $\mathfrak{L}^{\sim}$ to the category *Ab* of abelian groups such that the group $H^{\infty-n}(X; G)$ is isomorphic to the (n – 1)-th singular homology $H_{n-1}(E - X; G)$ of $E - X$.

As an immediate consequence we obtain the Alexander-Pontriagin Invariance Theorem in E: if the objects X and Y are equivalent or homotopically equivalent in $\mathfrak{L}^{\sim}$, then for each $n \geq 0$ the singular homology groups $H_n(E - X; G)$ and $H_n(E - Y; G)$ are isomorphic. A special case of this theorem (when $n = 0$ and $G = Z$) was proved for a complete E by J. Leray in [4].

We also note that the Leray-Schauder theory of topological degree follows readily from the special case of the main theorem.

---

* This report, presented by K. Geba under the title "The cohomology theory in Banach Spaces," outlines the results announced by the authors in [2].

§1. PRELIMINARIES

The directed set $\mathcal{L}$. Let E be an arbitrary but fixed infinite dimensional normed space. We shall denote by $\mathcal{L} = \{L_\alpha, L_\beta, L_\gamma, \ldots\}$ the directed set of all finite dimensional subspaces of E with the natural order relation $\leq$ defined by the condition

$$L_\alpha \leq L_\beta \Longleftrightarrow L_\alpha \subset L_\beta \ .$$

For notational convenience, we establish 1-1 correspondence $\alpha \Longleftrightarrow L_\alpha$ between the symbols $\alpha, \beta, \gamma, \ldots$ and $L_\alpha, L_\beta, L_\gamma, \ldots$ and in the formulas to occur we replace frequently one kind of symbols by another. Given $X \subset E$, we denote by $\mathcal{L}_X$ the cofinal subset of $\mathcal{L}$, consisting of those α for which $X_\alpha = X \cap L_\alpha$ is non-empty.

Compact fields. A mapping $F\colon X \to E$ from a metric space X is *compact* provided the closure of F(X) is compact. A compact mapping $F\colon X \to E$ is an *α-mapping* (resp. *finite dimensional mapping*) provided $F(X) \subset L_\alpha$ (resp. $F(X) \subset L_\beta$ for some β).

The following Schauder Approximation Theorem is of fundamental importance: every compact mapping can be uniformly approximated by finite dimensional mappings.

Given $X, Y \subset E$ and a mapping $f\colon X \to Y$, denote by the same but capital letter the mapping $F\colon X \to E$ defined by $F(x) = x - f(x)$ for $x \in X$. Then $f\colon X \to Y$ is said to be a *compact field* (resp. *α-field*) provided the corresponding $F\colon X \to E$ is a compact mapping (resp. α-mapping).

The subsets of E and the compact fields form a category, denoted by $\mathcal{C}$ and called the *Leray-Schauder category*. The subsets of E and the α-fields form the subcategory $\mathcal{C}_\alpha \subset \mathcal{C}$. The union of all categories $\mathcal{C}_\alpha$ is the subcategory $\mathcal{C}_0$ of $\mathcal{C}$. The morphisms of the category $\mathcal{C}_0$ will be called the *finite dimensional fields*.

Let $\mathcal{L}^\sim$ (resp. $\mathcal{L}_\alpha^\sim$ and $\mathcal{L}_0^\sim$) be the full subcategory of $\mathcal{C}$ (resp. $\mathcal{C}_\alpha$ and $\mathcal{C}_0$) generated by the closed and bounded subsets of E. The category $\mathcal{L}^\sim$ will be called the *main category*. By an object we shall understand in what follows an object of $\mathcal{L}^\sim$.

Compact homotopies. Given $X, Y \subset E$ and a homotopy $h_t: X \to Y$, $(0 \leq t \leq 1)$, denote by the capital H the mapping from $X \times I$ to E defined by $H(x, t) = x - h_t(x)$ for $(x, t) \in X \times I$. Then, $h_t: X \to Y$ is called a *compact homotopy* (resp. α-homotopy) provided $H: X \times I \to E$ is a compact mapping (resp. an α-mapping).

Two compact fields (resp. α-fields) $f_0, f_1: X \to Y$ are called *homotopic*, $f_0 \sim f_1$, (resp. *α-homotopic*, $f_0 \widetilde{\alpha} f_1$) provided there is a compact homotopy (resp. α-homotopy) $h_t: X \to Y$ such that $h_0 = f_0$, $h_1 = f_1$. Each of the above relations is clearly an equivalence relation. If the fields (resp. α-fields) $f_1, f_2: X \to Y$ and $g_1, g_2: Y \to Z$ are homotopic (resp. α-homotopic), then so are their compositions $g_1 \circ f_1, g_2 \circ f_2: X \to Z$.

It follows that the relations $\sim$ and $\widetilde{\alpha}$ convert the Leray-Schauder category $\mathcal{C}$ and the category $\mathcal{C}_\alpha$ into the h-categories $(\mathcal{C}, \sim)$ and $(\mathcal{C}_\alpha, \widetilde{\alpha})$ respectively. Similarly, we define the relation $\tilde{0}$ of finite dimensional homotopy and convert the category $\mathcal{C}_0$ into the h-category $(\mathcal{C}_0, \tilde{0})$.

It is evident that the main category $\mathcal{L}^{\sim}$ admits a structure of an h-category $(\mathcal{L}^{\sim}, \sim)$ with the relation of homotopy induced by that in $\mathcal{C}$. Similarly, the relation $\tilde{0}$ in $\mathcal{C}_0$ converts $\mathcal{L}_0^{\sim}$ into an h-category $(\mathcal{L}_0^{\sim}, \tilde{0})$.

An orientation in E. Denote by R^∞ the normed space consisting of all sequences $x = (x_1, x_2, \ldots)$ of real numbers such that $x_i = 0$ for all but a finite set of i, with the norm $\|x\| = \sqrt{\Sigma\, x_i^2}$ and put:

$$R^k = \{x \in R^\infty;\ x_i = 0 \text{ for } i \geq k+1\},$$

$$R_+^k = \{x \in R^k;\ x_k \geq 0\},$$

$$R_-^k = \{x \in R^k;\ x_k \leq 0\}.$$

Let α be an element of the directed set $\mathcal{L}$ and let $d(\alpha)$ be the dimension of α. Call two linear isomorphisms $l_1, l_2: L_\alpha \to R^{d(\alpha)}$ *equivalent*, $l_1 \sim l_2$, provided $l_1 l_2^{-1} \in GL_+(d(\alpha))$, i.e., the determinant of the corresponding matrix is positive. With respect to the relation $\sim$, the set of all linear isomorphisms from L_α to $R^{d(\alpha)}$ decomposes into exactly two equivalent classes. An arbitrary choice of one of these classes will be called an orientation of L_α.

Let us choose now an orientation $\mathcal{O}_\alpha$ of L_α for each α and call the family $\mathcal{O} = \{\mathcal{O}_\alpha\}$ an orientation in E. Given $\alpha < \beta$ with $d(\beta) - d(\alpha) = 1$ and $1_\alpha \in \mathcal{O}_\alpha$, there exists $1_\beta \in \mathcal{O}_\beta$ such that $1_\beta(x) = 1_\alpha(x)$ for all $x \in L_\alpha$.

We let

$$L_\beta^+ = 1_\beta^{-1}(R_+^{d(\beta)}) \quad \text{and} \quad \dot{L}_\beta^- = 1_\beta^{-1}(R_-^{d(\beta)}) \ .$$

Clearly, the definition of L_β^+ and L_β^- depends only on the orientation of L_α and L_β.

Given an object X we let

$$X_\beta^+ = X \cap L_\beta^+ \quad \text{and} \quad X_\beta^- = X \cap L_\beta^- \ .$$

If $f\colon X \to Y$ is an α-field, then clearly $f(X_\alpha) \subset Y_\alpha$ and hence f determines the map $f_\alpha\colon X_\alpha \to Y_\alpha$ given by $f_\alpha(x) = f(x)$ for $x \in X_\alpha$. We note that if α, β are two elements of $\mathcal{L}_X$ with $\alpha < \beta$, $d(\beta) - d(\alpha) = 1$, and $f\colon X \to Y$ is an α-field, then $f_\beta\colon X_\beta \to Y_\beta$ maps the triad $(X_\beta, X_\beta^+, X_\beta^-)$ into the triad $(Y_\beta, Y_\beta^+, Y_\beta^-)$.

§2. THE COHOMOLOGY FUNCTOR $\bar{H}^{\infty - n}\colon \tilde{\mathcal{L}}_0 \to Ab$.

Let G be an arbitrary but fixed Abelian group and n a positive integer. By H^* (resp. H_*) we shall denote the Cech cohomology (resp. the singular homology) with coefficients in G. By H^0 and H_0 we shall denote the reduced zero-dimensional groups.

A Lemma on the Alexander Pontriagin duality. For a compact subset $X \subset R^k$ we shall denote by D_k the Alexander-Pontriagin isomorphism

$$D_k\colon H_{n-1}(R^k - X) \to H^{k-n}(X)$$

determined by the standard orientation of R^k (see [5], p. 296). Let $X_0 = X \cap R^{k-1}$, $i\colon R^{k-1} - X_0 \to R^k - X$ be the inclusion and denote by Δ^* the Mayer-Vietoris homomorphism of the triad $(X, X \cap R_+^k, X \cap R_-^k)$. It follows from the definition of D_k that the following diagram is sign-commutative:

$$\begin{array}{ccc} H^{k-1-n}(X_0) & \xrightarrow{\Delta^*} & H^{k-n}(X) \\ \Big\downarrow D_{k-1} & & \Big\downarrow D_k \\ H_{n-1}(R^{k-1} - X_0) & \xrightarrow{i*} & H'_{n-1}(R^k - X) \end{array}$$

Thus, $D_k \circ \Delta = a_{k,n} i* \circ D_{k-1}$, where $a_{k,n} = \pm 1$. Let $b_{k,n}$, $k \geq n$, be defined by induction on k:

$$b_{n,n} = 1, \quad b_{k,n} = a_{k,n} b_{k-1,n} \quad .$$

Define the isomorphism

$$\mathcal{D}_k\colon \ H^{k-n}(X) \to H_{n-1}(R^k - X)$$

by $\mathcal{D}_k = b_{k,n} D_k'$.

We have the following:

LEMMA A. *Let* X *be a compact subset of* R^{k+1}, $X_0 = X \cap R^k$ *and let* Δ^* *be the Mayer-Vietoris homomorphism of the triad* $(X, X \cap R^{k+1}_+, X \cap R^{k+1}_-)$. *Then the following diagram commutes*

$$\begin{array}{ccc} H^{k-n}(X_0) & \xrightarrow{\Delta*} & H^{k-n+1} \\ \Big\downarrow \mathcal{D}_k & & \Big\downarrow \mathcal{D}_{k+1} \\ H_{n-1}(R^k - X_0) & \xrightarrow{i*} & H_{n-1}(R^{k+1} - X) \end{array}$$

(i: $R^k - X_0 \to R^{k+1} - X$ *denotes the inclusion*).

Let n be a positive integer, X be an object and $U = E - X$. Let $\mathcal{O} = \{\mathcal{O}_\alpha\}$ be an orientation in E and $1_\alpha \in \mathcal{O}_\alpha$.

For each $\alpha \in \mathcal{L}_X$ with $d(\alpha) > n$ define

$$\mathcal{D}_\alpha\colon \ H^{d(\alpha)-n}(X_\alpha) \to H_{n-1}(U_\alpha)$$

to be the Alexander-Pontriagin isomorphism "transferred" by 1_α^{-1} from $R^{d(\alpha)}$ to L_α.

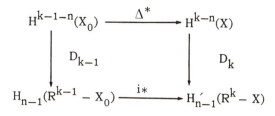

LEMMA A$'$. *Let* X *be an object,* $U = E - X$ *and* $\alpha, \beta \in \mathcal{L}_X$ *with* $d(\beta) - d(\alpha) = 1$. *Let* $i_{\alpha\beta}\colon U_\alpha \to U_\beta$ *be the inclusion and*

$$\Delta_{\alpha\beta}\colon H^{d(\alpha)-n}(X_\alpha) \to H^{d(\beta)-n}(X_\beta)$$

be the Mayer-Vietoris homomorphism of the triad (X, X^+, X^-). *Then the following diagram commutes*

$$\begin{array}{ccc} H^{d(\alpha)-n}(X_\alpha) & \xrightarrow{\ \Delta_{\alpha\beta}\ } & H^{d(\beta)-n}(X_\beta) \\ \Big\downarrow \mathfrak{D}_\alpha & & \Big\downarrow \mathfrak{D}_\beta \\ H_{n-1}(U_\alpha) & \xrightarrow{\ (i_{\alpha\beta})_*\ } & H_{n-1}(U_\beta) \end{array} .$$

This follows from the definition of $\Delta_{\alpha\beta}$ and Lemma A.

The group $H^{\infty-n}(X)$. Let $\alpha, \beta \in \mathcal{L}_X$, with $\alpha < \beta$, and let $\alpha = \alpha_0 < \alpha_1 < \cdots < \alpha_k = \beta$ be a chain of elements of $\mathcal{L}_X$ such that $d(\alpha_{i+1}) - d(\alpha_i) = 1$ $(i = 0, 1, \ldots, k-1)$. Define the homomorphism

$$\Delta_{\alpha\beta}\colon H^{d(\alpha)-n}(X_\alpha) \to H^{d(\beta)-n}(X_\beta)$$

to be the composition of the homomorphisms $\Delta_{\alpha\alpha_1}, \Delta_{\alpha_1\alpha_2}, \ldots, \Delta_{\alpha_{k-1}\beta}$. It follows from Lemma A$'$ that the definition of $\Delta_{\alpha\beta}$ does not depend on the choice of the chain $\alpha_1, \ldots, \alpha_{k-1}$ joining α and β.

For an object X consider the groups $H^{d(\alpha)-n}(X_\alpha)$ together with the homomorphisms $\Delta_{\alpha\beta}$ given for $\alpha \leq \beta$. The family $\{H^{d(\alpha)-n}(X_\alpha), \Delta_{\alpha\beta}\}$ indexed by $\alpha \in \mathcal{L}_X$ is a direct system of Abelian groups called the $(\infty-n)$-th *cohomology system of* X corresponding to the orientation $\mathcal{O}$ in E.

We define the Abelian group

$$H^{\infty-n}(X) = \varinjlim_{\alpha} \{H^{d(\alpha)-n}(X_\alpha), \Delta_{\alpha\beta}\}$$

to be the direct limit of the $(\infty-n)$-th cohomology system of X.

We note, taking into account Lemma A$'$, that if $\{H^{d(\alpha)-n}(X_\alpha), \Delta_{\alpha\beta}\}$, $\{H^{d(\alpha)-n}(X_\alpha), \overline{\Delta}_{\alpha\beta}\}$ are two $(\infty-n)$-th cohomology systems of X corresponding to different orientations $\mathcal{O} = \{\mathcal{O}_\alpha\}$ and $\overline{\mathcal{O}} = \{\overline{\mathcal{O}}_\alpha\}$ in E, then the above systems are isomorphic. Consequently, the group $H^{\infty-n}(X)$ does not depend on the orientation $\mathcal{O}$ in E.

f^* *for finite dimensional* f. Let $X, Y \in \mathcal{L}_0^{\sim}$, $\alpha_0 \in \mathcal{L}_X$ and $f: X \to Y$ be an α_0-field. Then for each $\alpha, \beta \in \mathcal{L}_X$ with $\alpha_0 \leq \alpha < \beta$ the following diagram commutes:

$$\begin{array}{ccc} H^{d(\alpha)-n}(X_\alpha) & \xleftarrow{f_\alpha^*} & H^{d(\alpha)-n}(Y_\alpha) \\ \downarrow \Delta_{\alpha\beta} & & \downarrow \Delta_{\alpha\beta} \\ H^{d(\beta)-n}(X_\beta) & \xleftarrow{f_\beta^*} & H^{d(\beta)-n}(Y_\beta) \end{array} .$$

Consequently, f induces a map $\{f_\alpha{}^*\}$ from the $(\infty-n)$-th cohomology system of Y into that of X, and therefore determines a map

$$f^* = \operatorname*{Lim}_{\substack{\to \\ \alpha}} \{f_\alpha{}^*\}$$

from $H^{\infty-n}(Y)$ to $H^{\infty-n}(X)$.

Let us denote now by $\overline{H}^{\infty-n}$ the functor which assigns to an object $X \in \mathcal{L}_0^{\sim}$ the $(\infty-n)$-th cohomology group $H^{\infty-n}(X)$ and to a finite-dimensional field $f: X \to Y$ the induced homomorphism $f^*: H^{\infty-n}(Y) \to H^{\infty-n}(X)$.

From the properties of the Mayer-Vietoris homomorphism we deduce the following:

THEOREM 1. *The functor* $\overline{H}^{\infty-n}$ *is a contravariant* h-*functor from the category* $\mathcal{L}_0^{\sim}$ *to the category of Abelian groups Ab.*

The continuity of the functor $\overline{H}^{\infty-n}$. In order to extend the functor $\overline{H}^{\infty-n}: \mathcal{L}_0^{\sim} \to Ab$ over $\mathcal{L}^{\sim}$, we shall use the property which is analogous to the property of continuity of the Cech cohomology.

We shall make use of the following algebraic lemma:

LEMMA B. *Let* $\mathfrak{N} = \{k, l, m, \ldots\}$ *and* $\mathcal{L} = \{\alpha, \beta, \gamma, \ldots\}$ *be two directed sets and* $\{H_\alpha^k, \cdot\}$ *be a double direct system of abelian groups indexed by* $\mathfrak{N} \times \mathcal{L}$. *Then we have a natural isomorphism between the limit groups*

$$\operatorname*{Lim}_{\vec{\alpha}} \operatorname*{Lim}_{\vec{k}} \{H_\alpha^k, \cdot\} \approx \operatorname*{Lim}_{\vec{k}} \operatorname*{Lim}_{\vec{\alpha}} \{H_\alpha^k, \cdot\} .$$

Let X be an object. For a natural number k let

$$X^{(k)} = \{x \in E;\ \rho(x, X) \leq \frac{1}{k}\} .$$

We shall say that a sequence of objects $\{X_k\}$ is an *approximating sequence* or an (a)-*sequence* for X provided

(i) $X_k \supset X_{k+1}$ for each $k = 1, 2, \ldots,$

(ii) $X = \bigcap_{k=1} X_k$.

We note that

(a) if $\{X_k\}$ is an (a)-sequence for X, then the enlarged sequence $\{\tilde{X}_k\}$, where $\tilde{X}_k = (X_k)^k = \{x \in E;\ \rho(x, X_k) \leq 1/k\}$ is also an (a)-sequence for X.

(b) if $\{X_k\}$ and $\{Y_k\}$ are two (a)-sequences for X and Y, respectively, then $\{X_k \cup Y_k\}$ is an (a)-sequence for $X \cup Y$.

(c) if $\{X_k\}$ is an (a)-sequence for X and $f\colon X_1 \to E$ is a compact field, then $\{f(X_k)\}$ is an (a)-sequence for $f(X)$.

Let Y be an object, $\alpha \in \mathcal{L}_Y$ and let $\{Y_k\}$ be an (a)-sequence for Y. We let $Y_\alpha^k = Y_k \cap L_\alpha$. Consider the following inclusions, all of them being finite-dimensional fields:

$$i_{kl}\colon Y_l \to Y_k,\ i_{kl}^\alpha\ Y_\alpha^l \to Y_\alpha^k, \qquad (k \leq l) ;$$

$$j_k\colon Y \to Y_k,\ j_{k\alpha}\colon Y \to Y^k,$$

and the direct system of Abelian groups $\{H^{\infty-n}(Y_k), i_{kl}^*\}$ over $\mathfrak{N}$. We have the commutativity relations $j_k^* = j_l^* \circ i_{kl}^*$ for $k \leq l$ and hence

$$\{j_k^*\}\colon \{H^{\infty-n}(Y_k), i_{kl}^*\} \to H^{\infty-n}(Y)$$

is a direct family of homomorphisms.

THEOREM 2. *The map*

$$\operatorname*{Lim}_{\vec{k}} \{j_k^*\}\colon \operatorname*{Lim}_{\vec{k}} \{H^{\infty-n}(Y_k),\, i_{k1}^*\} \to H^{\infty-n}(Y)$$

is an isomorphism.

Proof: Consider over $\mathfrak{N} \times \mathfrak{L}$ the following double direct systems of Abelian groups $\kappa = \{H^{d(\alpha)-n}(Y_\alpha^k), \Delta_{\alpha\beta}^{k1}\}$, and $\bar{\kappa} = \{H^{d(\alpha)-n}(Y_\alpha), \Delta_{\alpha\beta}\}$, where $\Delta_{\alpha\beta}^{k1} = (i_{k1}^*) \circ \Delta_{\alpha\beta}$, for $\alpha \leq \beta$, $k \leq 1$. Clearly $\{j_{k\alpha}^*\}$ is a map from κ to $\bar{\kappa}$.

In view of the continuity of the Cech cohomology [1], the map

$$\operatorname*{Lim}_{\vec{k}} \{j_{k\alpha}\}$$

is an isomorphism for each α, and therefore so is the map

$$\operatorname*{Lim}_{\vec{\alpha}} \operatorname*{Lim}_{\vec{k}} \{j_{k\alpha}^*\}.$$

Consequently, in view of the naturality of the isomorphism in Lemma B, the map

$$\operatorname*{Lim}_{\vec{k}} \{j_k^*\} = \operatorname*{Lim}_{\vec{k}} \operatorname*{Lim}_{\vec{\alpha}} \{j_{k\alpha}^*\}$$

is also an isomorphism and the proof is completed.

§4. THE FUNCTOR $H^{\infty-n}$: $\mathfrak{L}^{\sim} \to Ab$.

Let X, Y be two objects and let $f\colon X \to Y$ be a compact field. A sequence $\{Y_k, f_k\}$ of objects Y_k and α_k-fields $f_k\colon X \to Y_k$ is called an *approximating system* or (a) system for f, provided:

(i) $\{Y_k\}$ is an (a)-sequence for Y,

(ii) $f_k \sim j_k f$ in $\mathfrak{L}^{\sim}$, where $j_k\colon Y \to Y_k$ is the inclusion,

(iii) $f_k \sim i_{k1} f_1$ in $\mathfrak{L}_0^{\sim}$, where $i_{k1}\colon Y_1 \to Y_k$ is the inclusion $(k \leq 1)$.

Every compact field $f\colon X \to Y$ admits an (a)-system. In fact, let $Y_k = Y^{(k)}$, and let $f_k\colon X \to Y_k$ be an α_k-field such that

$$\|f(x) - f_k(x)\| \leq \frac{1}{k} \qquad \text{for all } x \in X.$$

Clearly, $\{Y_k, f_k\}$ is an (a)-system for f. In what follows any such system $\{Y_k, f_k\}$ will be called a standard (a)-system for f.

Let $f\colon X \to Y$ be a compact field and $\{Y_k, f_k\}$ be an arbitrary (a)-system for f. In view of (iii) the diagram

$$\begin{array}{ccc} H^{\infty-n}(Y_k) & \xrightarrow{\;i_{kl}^*\;} & H^{\infty-n}(Y_l) \\ & {\scriptstyle f_k}\searrow \quad \swarrow{\scriptstyle f_l} & \\ & H^{\infty-n}(X) & \end{array}$$

commutes in *Ab*. Consequently, $\{f_k^*\}$ is a direct family of maps and

$$\underset{\vec{k}}{\operatorname{Lim}}\, \{f_k^*\}\colon\ \underset{\vec{k}}{\operatorname{Lim}}\, \{H^{\infty-n}(Y_k),\ i_{kl}^*\} \to H^{\infty-n}(X)\ .$$

We define the induced homomorphism

$$f^*\colon\ H^{\infty-n}(Y) \to H^{\infty-n}(X)$$

by the formula

$$f^* = \underset{\vec{k}}{\operatorname{Lim}}\, \{f_k^*\} \circ (\underset{\vec{k}}{\operatorname{Lim}}\, \{j_k^*\})^{-1}\ .$$

It is easily shown that the definition of f^* does not depend on the choice of $\{Y_k, f_k\}$.

Now define the functor $H^{\infty-n}$ from the main category $\mathcal{L}^{\sim}$ to the category of Abelian groups *Ab* by putting $H^{\infty-n}(X) = \overline{H}^{\infty-n}(X)$, $H^{\infty-n}(f) = f^*$. If $f\colon X \to Y$ is an α-field, it is easily seen (by taking $\{Y_k, f_k\}$ with $Y_k = Y$, $f_k = f$) that $\overline{H}^{\infty-n}(f) = H^{\infty-n}(f)$, i.e., $H^{\infty-n}$ extends $\overline{H}^{\infty-n}$ over $\mathcal{L}^{\sim}$.

THEOREM 3. *The induced map* f^* *satisfies the following properties:*

(a) *the homotopy* $f \sim g$ *implies* $f^* = g^*$;

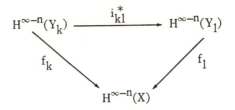

(b) $(gf)^* = f^* \circ g^*$.

In other words, $H^{\infty-n}$ *is an* h*-functor from the main category* $\mathcal{L}^{\sim}$ *to the category of Abelian groups* Ab.

Proof of the property (a). Let f, g: $X \to Y$ be two compact fields and h_t: $X \to Y$ be a compact homotopy such that $f = h_0$, $g = h_1$. Let $h_t^{(k)}$: $X \to Y^{(k)}$ be an α_k-homotopy satisfying

$$\|h_t^{(k)}(x) - h_t(x)\| \le \frac{1}{k} \quad \text{for all } (x, t) \in X \times I.$$

Let $f_k = h_0^{(k)}$, $g_k = h_1^{(k)}$. Evidently, $\{Y^{(k)}, f_k\}$ and $\{Y^{(k)}, g_k\}$ are (a)-systems for f and g respectively. Thus, since $f_k \underset{\alpha_k}{\sim} g_k$, we have

$$\operatorname*{Lim}_{\vec{k}} \{f_k^*\} = \operatorname*{Lim}_{\vec{k}} \{g_k^*\}$$

and therefore $f^* = g^*$.

Proof of the property (b). (Special Case). Let f: $X \to Y$ be a compact field, g: $Y \to Z$ be an α_0-field and let $h = gf$. We shall prove that $h^* = f^* \circ g^*$.

Take a standard (a)-system $\{Y^{(k)}, f_k\}$ for f and take an arbitrary finite dimensional extension $\bar{g}$: $Y^{(1)} \to E$ of g over $Y^{(1)}$. The existence of g follows from the Tietze Extension Theorem.

Let us put $W_k = \bar{g}(Y^{(k)}) \cup Z$ for each k and consider the enlarged (a)-sequence $\{\tilde{W}_k\}$ for Z.

In the diagram

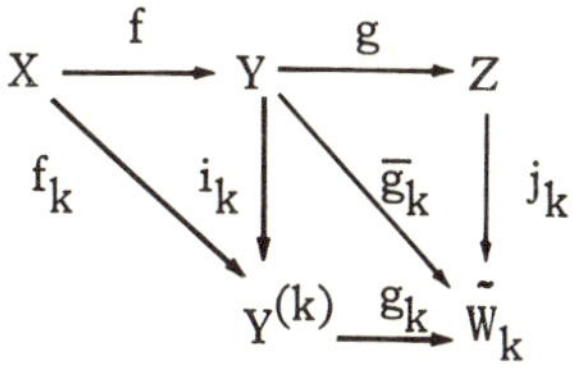

define g_k by putting $g_k(y) = \bar{g}(y)$ for all $y \in Y^{(k)}$ and let $\bar{g}_k = g_k i_k$, $h_k = g_k f_k$. It is easily seen that $\{\tilde{W}_k, h_k\}$, $\{\tilde{W}_k, g_k\}$ are (a)-systems for h and g, respectively. We have

$$\operatorname*{Lim}_{\vec{k}} \{h_k^*\} = \operatorname*{Lim}_{\vec{k}} \{f_k^*\} \operatorname*{Lim}_{\vec{k}} \{g_k^*\} = \operatorname*{Lim}_{\vec{k}} \{f_k^*\} (\operatorname*{Lim}_{\vec{k}} \{i_k^*\})^{-1} \operatorname*{Lim}_{\vec{k}} \{\bar{g}_k^*\}$$

and thus

$$\operatorname*{Lim}_{\vec{k}} \{h_k^*\} = f^*(\operatorname*{Lim}_{\vec{k}} \{\bar{g}_k^*\}) .$$

This implies $h^* = f^* \circ g^*$, and the proof is completed.

Proof of the property (b). (General case) Let $f\colon X \to Y$ and $g\colon Y \to Z$ be two compact fields and let $h = gf$.

Let $\{Z^{(k)}, h_k\}$ and $\{Z^{(k)}, g_k\}$ be two standard (a)-systems for h and g, respectively.

From the inequalities

$$\|h_k(x) - h(x)\| \leq \frac{1}{k} \quad \text{for all } x \in X,$$

$$\|g_k f(x) - h(x)\| \leq \frac{1}{k} \quad \text{for all } x \in X,$$

it follows that the fields $g_k f, h_k\colon X \to Z^{(k)}$ are homotopic in $\mathfrak{L}^{\sim}$ for each $k = 1, 2, \ldots$. This implies, in view of the property (a), that $h_k^* = (g_k f)^*$. Since each g_k is finite-dimensional, we have $(g_k f)^* = f^* g_k^*$ and thus we obtain

$$\operatorname*{Lim}_{\vec{k}} \{h_k^*\} = \operatorname*{Lim}_{\vec{k}} \{f^* g_k^*\} = f^* \operatorname*{Lim}_{\vec{k}} \{g_k^*\} \cdot$$

This implies $h^* = f^* g^*$ and the proof of Theorem 3 is completed.

§5. THE ALEXANDER-PONTRIAGIN DUALITY IN E.

In view of Lemma A′ the $(\infty - n)$-th cohomology system of an object X is isomorphic to the direct system of homology groups $\{H_{n-1}(U_\alpha), (i_{\alpha\beta})_*\}$, where $U = E - X$. We have clearly

$$H_{n-1}(U) \cong \operatorname*{Lim}_{\vec{\alpha}} \{H_{n-1}(U_\alpha), (i_{\alpha\beta})_*\} .$$

Consequently, in view of the definition of the group $H^{\infty - n}(X)$, we obtain the following theorem:

THEOREM 4. *For every object* X *we have an isomorphism*

$$H^{\infty-n}(X) \cong H_{n-1}(E - X)$$

between the $(\infty - n)$*-th cohomology group of* X *and the* $(n-1)$*-th singular homology group of the complement of* X *in* E.

Two objects X and Y are called equivalent (resp. h-*equivalent*) in $\mathfrak{L}^{\sim}$ provided there exist compact fields $f: X \to Y$, $g: Y \to X$ such that $gf = 1_X$, $fg = 1_Y$ (resp. $gf \sim 1_X$, $fg \sim 1_Y$).

It follows from Theorem 3 that if the objects X and Y are equivalent or h-equivalent in $\mathfrak{L}^{\sim}$. then the $(\infty - n)$-th cohomology groups $H^{\infty - n}(X; G)$ and $H^{\infty - n}(Y; G)$ with the coefficients in G are isomorphic. From this, taking into account Theorem 4, we obtain the following.

THEOREM 5. *If* X *and* Y *are two equivalent or* h*-equivalent objects of the category* $\mathfrak{L}^{\sim}$, *then for each* $n = 0, 1, 2, \ldots$ *and a coefficient group* G, *the singular homology groups* $H_n(E - X; G)$ *and* $H_n(E - Y; G)$ *are isomorphic.*

§6. THE FUNCTOR $H^{\infty-1}$ AND THE LERAY-SCHAUDER DEGREE

Let $A(r, R) = \{x \in E;\ r \leq \|x\| \leq R\}$, $0 < r \leq R$. By Theorem 4, $H^{\infty-1}(A(r, R); Z) \approx Z$. We will identify groups $H^{\infty-1}(A(r, R); Z)$ with Z by isomorphisms which are compatible with the isomorphisms induced by the inclusions $A(r, R) \subset A(r_1, R_1)$, for $r \geq r_1$, $R \leq R_1$. We denote by 1 the generator of Z.

Let X be an object, $\{U_i\}_{i \in I}$ the family of all bounded components of $E - X$. Let $f_i(x) = x - u_i$, $u_i \in U_i$; $f_i: X \to A(r, R)$ for some $R \geq r > 0$. Evidently, f_i does not depend on u_i.

As a consequence of Theorem 4, we have:

THEOREM 6. $H^{\infty-1}(X; Z)$ *is a free Abelian group with generators* $\alpha_i = f_i^*(1)$. *If* $g: X \to A(r, R)$ *is a compact field, then* $g^*(1) = \Sigma_{q=1}^{p} n_q \alpha_{i_q}$.

The integer $n_q = \gamma(g, U_{i_q})$ is called the *Leray-Schauder index of* g *with respect to* U_{i_q}.

As an immediate consequence we obtain the main theorem of the Leray-Schauder theory of topological degree:

COROLLARY. *Let* U *be a bounded open set with boundary* X. *To every compact field* $f: \bar{U} \to E$ *and every point* $y \in E - f(X)$ *corresponds an integer* $d(y, f, U)$ *with the properties:*

(i) *if* $h_t: (\bar{U}, X) \to (E, E - y)$ *is a compact homotopy, then*

$$d(y, h_1, U) = d(y, h_0, U) ;$$

(ii) *if* $d(y, f, U) \neq 0$, *then* $y = f(x)$ *for some* $x \in U$;

(iii) *if* $U = \bigcup U_i$ *is the union of open disjoint sets* U_i *all having their boundary in* X, *then*

$$d(y, f, U) = \sum_i d(y, f, U_i) ,$$

(iv) *if* $A \subset \bar{U}$ *is closed and* $y \notin f(A)$, *then*

$$d(y, f, U) = d(y, f, U - A) .$$

If U is connected, consider the field $g: X \to A(r, R)$ defined by $g(x) = f(X) - y$ for $x \in X$ and put

$$d(f, U, y) = \gamma(g, U) .$$

Now, Theorem 6 permits to define $d(y, f, U)$ in an evident manner for an arbitrary U. The properties (i)-(iv) follow from Theorems 5 and 6.

BIBLIOGRAPHY

[1] S. Eilenberg and N. Steenrod, *Foundations of Algebraic Topology*, Princeton, 1952.

[2] K. Geba and A. Granas, Algebraic topology in linear normed spaces, *Bull. Acad. Polon. Sci.*, *I*, 13 (1965), pp. 287-290; *II*, 13 (1965) pp. 341-345; *III*, 15 (1967), pp. 137-143; *IV*, 15 (1967) pp. 145-152.

[3] J. Leray, Topologie des espaces abstraits de M. Banach, *C. R. Acad. Sci.*, Paris 200 (1935), pp. 1083-1093.

[4] L. Leray and J. Schauder, Topologie et équations fonctionnelles, *Ann. Ecole Norm. Sup.*, 51 (1934), pp. 45-78.

[5] E. Spanier, *Algebraic Topology*, New York, 1966.

Normal School, Gdańsk

Institute of Mathematics,

Polish Academy of Sciences

ANALYSE DE LA TECHNIQUE DE NASH-MOSER

par G. Glaeser (Rennes)(*)

I. *Introduction*

Le théorème des fonctions implicites concernant les fonctions de plusieurs variables s'étend aisément à des opérateurs definis dans un ouvert Ω d'un espace de Banach E.

Cette généralisation "ne coute pas cher." Mais son utilité, en Analyse et en Mathématiques Appliquées est assez réduite (pour des raisons que nous exposerons plus loin (cf. §II.E)).

En 1956, traitant du problème du plongement isométrique d'une variété Riemannienne dans un espace euclidien de dimension supérieure, John Nash a imaginé un raisonnement qui semblait recouvrir une généralisation du théorème élémentaire. On a reconnu que ces constructions présentaient des analogies avec des méthodes utilisées par Kolmogoroff et Arnold, pour résoudre des problèmes de Mecanique Céleste.

Jacob Schwartz puis Moser ont alors contribue à dégager les idées directrices qui se dissimulent derrière les cascades de majorations de Nash. C'est le premier pas vers l'énoncé d'un théorème des fonctions implicites appelé à rendre de grand service, dès qu'il sera énoncé d'une façon simple, c'est à dire abstraite! Mais ces écuries d'Augias sont loin d'être néttoyées à fond.

Nous allons d'abord commenter le théorème classique (sous sa variante dite "du difféomorphisme local") pour mettre en lumière les différences avec la situation rencontrée par Nash et Moser.

(*) L'élaboration de cet article a été facilitée par des discussions avec A. Brunel (Rennes) que je tiens a remercier ici.

II. *Le théorème du difféomorphisme local*

Soit Ω un voisinage ouvert de l'origine 0, dans un espace de Banach E, et soit θ un opérateur non linéaire défini dans Ω, à valeurs dans un Banach F. (On suppose $\theta(0) = 0$, pour simplifier.)

THÉORÈME: Si θ est de classe C^m (avec $m \geq 1$) et si sa différentielle à l'origine $D_0\theta \in \mathcal{L}(E,F)$ est un *isomorphisme* de E sur F, il existe un voisinage V de l'origine, contenu dans Ω tel que la restriction de θ à V soit un difféomorphisme.

(La démonstration complète se trouve dans les bons ouvrages.)

Commentaire:

A) On sait (Théorème de Banach) que $D_0\theta$ admet un inverse linéaire continu $L_0 \in \mathcal{L}(F,E)$.

B) Il s'agit essentiellement de résoudre, pour $f \in F$ suffisamment voisin de l'origine, l'équation $\theta(e) - f = 0$ (ou encore $\theta(e) = 0$, en introduisant l'opération

$$\theta : \theta(e) = \Theta(e) - f)) \, .$$

On utilise pour cela une méthode d'approximation successive. Partant d'un $u_0 = e_0$, suffisamment voisin de l'origine, dans Ω, on construit e sous forme d'une somme de série $e = \sum_0^\infty u_i$ normalement convergente. (La série des normes est majorée par une série géométrique.) Posons $e_n = \sum_0^n u_i$.

La "boucle de calcul" de l'algorithme est la suivante:

$$E \to F \to E \to E$$

$$e_n \mapsto \theta(e_n) \mapsto u_{n+1} = -L_0[\theta(e_n)] \mapsto e_{n+1} = e_n + u_{n+1} \, .$$

C) On remarque que le retour de F vers E se fait à l'aide de l'opérateur fixe L_0.

On peut (ce n'est pas indispensable dans le cas classique, mais essentiel dans la version Moser) accelerer la convergence en effectuant le retour grâce à l'inverse de la différentielle $D_{e_n}\theta$ en e_n (et non en 0).

C'est la *méthode de Newton.*

Au lieu d'obtenir une suite de u_n satisfaisant à $\|u_{n+1}\|_E \leq k\|u_n\|_E$ (avec $k < 1$), on aboutit à la décroissance accélerée

$$\|u_{n+1}\|_E \leq k\,\|u_n\|_E^2 \ .$$

Ceci ne peut s'obtenir que si θ est de classe C^m avec $m \geq 2$.

D) La démonstration du théorème classique utilise une *"boule de sécurité"* $\mathcal{B}$: c'est un voisinage de l'origine, contenu dans Ω, dans lequel les données satisfont à de bonnes majorations. Ces conditions assurent que, partant d'un $e_n \in \mathcal{B}$ on obtiendra, à la boucle suivante un $e_{n+1} \in \mathcal{B}$ ce qui permettra de recommencer sans s'échapper de l'ensemble Ω.

E) Supposons que E et F soient des espaces de fonctions de classe C^r sur une variété compacte.

Partant d'un $f \in F$ de classe C^r, on aboutit à un $e \in E$, qui est une fonction de classe C^r.

Or, l'expérience montre que dans de nombreux problèmes, partant de données régulières, on obtient des solutions moins régulières. Ainsi, chaque fois qu'un problème conduit à ce phénomène de perte de dérivabilité, on peut être assuré que le théorème élémentaire des fonctions implicites (employé sans modification) est inapte à le résoudre.

III. *Examen de la situation:*

A) Dans le cas de Moser, interviennent au moins 6 espaces de Banach*

$$M_1 \supset M \supset E_1 \supset E \quad \text{et} \quad F \supset \Phi \ .$$

Chaque espace est dense dans celui qui le contient et les injections sont continues.

L'opérateur non linéaire Θ (avec $\Theta(0) = 0$) défini dans un ouvert $\Omega \subset M$, à valeurs dans F est de classe $C^m(\Omega,F)$ $(m \geq 2)$ (plus précise-

---

* En fait, il y en a un sixième, intermédiaire entre M_1 et M, mais il nous semble, en premier examen, que l'on peut s'en passer.

ment, il doit être tel que les hypothèses que nous énumererons au paragraphe IV-D doivent être satisfaites). La restriction $\hat{\Theta}$ de Θ à $\Omega \cap E$ prend ses valeurs dans Φ. $\hat{\Theta}$ est également de classe C^m. Le diagramme (où les flèches horizontales représente des injections)

$$\begin{array}{ccc} \Omega \cap M & \longrightarrow & \Omega \\ \hat{\theta} \downarrow & & \downarrow \Theta \\ \Phi & \longrightarrow & F \end{array}$$

est commutatif. Diagramme analogue pour les diffentielles.

B) Pour suggerer l'usage que nous comptons faire de ces 6 espaces, indiquons dès maintenant leur interprétation dans le problème du plongement isométrique d'un tore T à deux dimensions muni d'une structure riemannienne, dans l'espace euclidien à 5 dimensions.

M_1, M, E_1, E; F, Φ, seront respectivement les espaces de fonctions définis sur le tore de classe:

$$C^{r-2},\ C^{r},\ C^{r+40},\ C^{r+42};\ C^{r-2},\ C^{r+40}\ .$$

Nous dirons que E, E_1 et Φ sont les espaces *lisses*; M_1, M et F les espaces *rugueux*.

Les *calculs* s'effectueront exclusivement dans les espaces E, E_1 et Φ. Par contre, la norme qui servira à faire la *majoration* définitive sera celle de M.

Ainsi, la convergence de la série (formée de termes $\in E$) que conduira à la réponse e s'effectuera dans l'espace M. Partant de données appartenant à Φ et E, on aboutit à une réponse $\in M$.

La perte de dérivabilité est de 42.

C) La différentielle $D_A\hat{\Theta} \in \mathcal{L}(E,\Phi)$ (pour $A \in \Omega \cap E$) est injective (*), mais son image $\operatorname{Im} D_A\hat{\Theta}$ est un sous-espace partout dense de Φ (distinct de Φ).

(*) En fait, dans l'exemple-pionnier de Nash, elle ne l'est pas. Mais ce n'est pas là la difficulté essentielle.

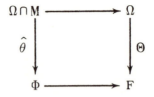

$D_A\hat{\Theta}$ ne possède par conséquent pas d'inverse à droite (de Φ dans E), et elle admet une infinité d'inverse à gauche, discontinues: il est impossible d'appliquer le théorème de Banach.

On se console en supposant l'existence d'un opérateur linéaire $\hat{L}_A \in \mathcal{L}(\Phi, E_1)$ tel que $\hat{L}_A \circ D_A\hat{\Theta}$ soit l'identité dans E. Mais $\hat{L}_A$ peut aussi s'appliquer à des éléments de $\Phi - \operatorname{Im} D_A\hat{\Theta}$, il les envoie dans $E_1 \supset E$.

Ainsi on arrive à inverser $D_A\hat{\Theta}$ au prix d'une perte de dérivabilité (dans l'exemple invoqué elle est de $2 = 42 - 40$).

D) Nous supposerons en outre qu'il existe $L_A \in \mathcal{L}(F, M)$ tel que $L_A \circ D_A\Theta$ soit l'identité de M.

Posant $\theta(x) = \Theta(x) - f$ (et $\widehat{\theta(x)} = \hat{\theta}(x) - f$) on aboutit au diagramme commutatif (les verticales sont les injections)

$$\begin{array}{ccccc} E & \xrightarrow{\hat{\theta}} & \Phi & \xrightarrow{\hat{L}_A} & E_1 \\ \downarrow & & \downarrow & & \downarrow \\ M & \xrightarrow{\theta} & F & \xrightarrow{L_A} & M_1 \end{array}$$

E) Si l'on tentait d'utiliser l'algorithme décrit en II, B, on perdrait de la derivabilité à chaque boucle. Au bout d'un nombre fini, d'itérations toutes les réserves seraient épuisées.

IV. *Analyse des techniques*

A) Pour obvier à cet inconvénient Nash préconise la boucle de calcul suivante:

$$E \xrightarrow{\hat{\theta}} \Phi \xrightarrow{\hat{L}_{e_n}} E_1 \xrightarrow{T(n)} E \longrightarrow E$$

$$e_n \to \widehat{\theta(e_n)} \to v_n = -\hat{L}_{e_n}[\widehat{\theta(e_n)}] \to u_{n+1} \to e_{n+1} = e_n + u_{n+1} .$$

Le passage de E_1 à E se fait grâce à une suite *d'opérateurs de lissage* T(n), qui détruisent les pertes de derivabilité à chaque boucle.

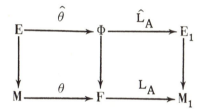

B) Ces opérateurs de lissage seront dans la pratique, des opérateurs de convolutions. Mais leur choix devra être très ingénieux pour que le mécanisme fonctionne (cf. Bonic [1]).

(Signalons, que dans l'exemple du plongement isométrique, Moser utilise l'opérateur qui associe à une fonction de classe C^r, périodique, la somme des $N(n)$ premiers termes de la séries de Fourier (convolution par un noyau de Dirichlet), où les $N(n)$ sont des entiers qui croissent d'une facon convenable avec n).

Un opérateur de lissage T appartient à $\mathcal{L}(M_1,E)$. (Défini dans l'espace le plus rugueux, il prend ses valeurs dans l'espace le plus lisse.)

On peut, en le composant avec des injections canoniques l'identifier à un élément de $\mathcal{L}(A,B)$ où A et B sont deux des espaces de Banach

$$M_1 \supset M \supset E_1 \supset E .$$

Adoptons le symbole $|||T|||_A^B$ pour désigner la norme de T dans l'espace de Banach $\mathcal{L}(A,B)$.

La propriété la plus banale que l'on exige d'une suite d'opérateurs de lissage est la suivante: Si I_A désigne l'application identique de A sur A et si B est plus lisse que A, la suite des $|||I_A - T(n)|||_B^A$ doit tendre vers 0 avec $1/n$; nous aurons plus particulièrement à étudier la fonction:

$$\frac{1}{\alpha(n)} = |||I_{M_1} - T(n)|||_{E_1}^{M_1} .$$

Par contre, si B est plus lisse que A la suite $|||T(n)|||_A^B$ n'est, en général, pas bornée. Nous aurons plus particulièrement à étudier les fonctions:

$$\beta(n) = |||T(n)|||_{M_1}^{M} \qquad \beta'(n) = |||T(n)|||_{E_1}^{E} .$$

Les propriétés requises par la suite des opérateurs de lissage pour que l'algorithme puisse fonctionner concernent la croissance des fonctions $\alpha(n)$, $\beta(n)$ (et aussi $\gamma(n)$ et $\delta(n)$ dont il sera question plus loin).

Il faut que $\alpha(n)$ croisse suffisamment vite pour compenser la croissance des autres fonctions. On y parvient en creusant un écart considérable (42 dans le cas étudié) entre la lissité de M_1 et E_1 (ce qui conduit à une très bonne approximation de I_{M_1} par les T(n)), en comparaison de l'ecart relativement faible (ici 2) entre la lissite de M et M_1 (resp. E et E_1).

C) Pour s'assurer que les e_n ne s'échapperont pas en cours de calcul, de l'ouvert Ω, on les enferme dans une "boule de sécurité" $\mathcal{B}$. Ce sera une M-boule de centre $u_0 = e_0$, dont le rayon 1/a sera précisé bientôt. (On doit avoir $\mathcal{B} \subset \Omega$.) $\mathcal{B}$ se definit par $\|x-u_0\|_M \leq 1/a$. Les e_n devront appartenir à $\mathcal{B} \cap E$ qui n'est pas une E-boule: et l'on ne pourra obliger la fonction

$$\gamma(n) = \|e_n - u_0\|_E$$

à rester bornée.

Une des finesses de la méthode consistera à controler cette croissance, grâce à un choix judicieux de T(n).

Esquissons un calcul de majoration de $\gamma(n)$

$$e_{n+1} = e_n + u_{n+1} \quad \text{et} \quad u_{n+1} = T_n(v_n)$$

$$\gamma(n+1) \leq \|u_{n+1}\|_E + \gamma(n) \leq \beta(n) \|v_n\|_{E_1} + \gamma(n) .$$

Mais $v_n = -\hat{L}_{e_n}[\hat{\theta}(e_n)]$. Anticipant sur la formule III-(D)-5 ultérieure nous en déduirons

$$\gamma(n+1) \leq (a^2 \beta'(n) + 1) \gamma(n)$$

ce qui montre que l'on peut majorer γ à l'aide de la seule fonction β'.

D) *Enonçons les hypothèses* que nous exigeons de la fonction θ, (et de L).

Nous supposons d'abord que les applications $x \to D_x\theta$ (resp. $x \to D_x\hat{\theta}$, resp. $x \to L_x$, resp. $x \in \hat{L}_x$) définis dans Ω, et à valeurs dans $\mathcal{L}(M,F)$, reps. $\mathcal{L}(E,\Phi)$, reps. $\mathcal{L}(F,M_1)$, resp. $\mathcal{L}(\Phi,E_1)$ soeint continues et localement bornées. Cela permettra de choisir un nombre a assez grand pour que, en tout point x de la M-boule (de rayon 1/a) on ait

(III-D-1) $$|||L_x|||_F^{M_1} \leq a \qquad |||\hat{L}_x|||_\Phi^{E_1} \leq a$$

2) $$|||D_x\theta|||_M^F \leq a \qquad |||D_x\hat{\theta}|||_E^\Phi \leq a .$$

De plus nous postulons la validité des inégalités tayloriennes suivantes:

$$\forall x \in \mathcal{B} \quad \text{et} \quad x+W \in \mathcal{B} \quad \text{on a}$$

3) $$\|\theta(x+w) - \theta(x) - D_x\theta\,[w]\|_F \leq a\|w\|_M^2 .$$

Enfin $$\|\widehat{\theta(x)}\|_\Phi \leq \|\widehat{\theta(u_0)}\|_\Phi + a\|x - u_0\|_E .$$

Cette dernière inégalité ne nous est utile que dans le cas où $\|x - u_0\|_E$ est très grand (cf. la majoration de $\gamma(n)$ au § précédent); comme $\|\widehat{\theta(u_0)}\|_\Phi$ est une constante, on pourra (en augmentant au besoin le nombre a) remplacer cette inégalité par

D (4) $$\|\widehat{\theta(x)}\|_\Phi \leq a\|x - u_0\|_E$$

qui combinée avec (1) donne

D (5) $$\|\hat{L}_x[\widehat{\theta(x)}]\|_{E_1} \leq a^2\|x - u_0\|_E .$$

E) *La majoration*: on trouve successivement

E (1) $$\|u_{n+1}\|_M = \|T(n)\,v_n\|_M \leq \beta(n)\|v_n\|_{M_1}$$

E (2) $$\|v_n\|_{M_1} = \|L_{e_n}\theta(e_n)\|_{M_1} \leq a\|\theta(e_n)\|_F .$$

Appliquons (D-3) avec $x = e_{n-1}$ et $w = u_n$ (donc $x+w = e_n$), on trouve

$$\|\theta(e_n)\|_F \leq \|\theta(e_{n-1}) + D_{e_{n-1}}\theta[u_n]\|_F + a\|u_n\|_M^2$$

Remarquons que par définition de V_{n-1}, on a $\theta(e_{n-1}) + D_{e_{n-1}}\theta[V_{n-1}] = 0$ (pour s'en assurer composer les deux membres avec $L_{e_{n-1}}$). On aboutit à $V_{n-1} = -L_{e_{n-1}}[\theta(e_{n-1})]$. Il en resulte que

$$\|\theta(e_n)\|_F \leq \|D_{e_{n-1}}\theta[u_n - V_{n-1}]\|_F + a\|u_n\|_M^2 \leq a\|u_n - V_{n-1}\|_M + a\|u_n\|_M^2 .$$

Mais $u_n = T(n)\,[v_{n-1}]$. Donc

$$\|\theta(e_n)\|_F \leq \frac{a}{\alpha(n)}\|v_{n-1}\|_{E_1} + a\|u_n\|_M^2 .$$

D'après (D-5): $\|v_{n-1}\|_{E_1} = \|\hat{L}_{e_{n-1}}[\widehat{\theta(e_{n-1})}]\|_{E_1} \leq a^2\,\gamma(n-1)$.

D'où

E (3) $$\|\theta(e_n)\|_F \leq \frac{a^3\, \gamma(n-1)}{\alpha(n)} + a\|u_n\|_M^2 \,.$$

On aboutit finalement à l'inégalité

E (4) $$\|u_{n+1}\|_M \leq a^4 \frac{\beta(n)\, \gamma(n-1)}{\alpha(n)} + a^2\, \beta(n) \cdot \|u_n\|_M^2 \,.$$

F) *L'astuce finale*: La formule E_4 laisse peu d'espoir au premier examen, car le dernier terme contient l'infiniment grand $\beta(n)$.

On peut souhaiter que la décroissance escomptée de $\|u_n\|_M^2$ compense la croissance de $\beta(n)$. J. Moser remarque alors, que dans l'exemple qu'il étudie, les fonctions α, β, γ pourront être choisie (grâce à un choix judicieux des $T(n)$) de façon à ce qu'il existe une fonction δ croissante satisfaisant aux deux conditions suivantes:

F (1) $$\beta(n) \cdot \delta(n+1) \leq (\delta(n))^2 \,.$$

(Cette condition impose à δ de croître plus vite que β.)

F (2) La série de terme général

$$\frac{\beta(n)\, \gamma(n-1)\, \delta(n+1)}{\alpha(n)} \quad \text{doit être convergente.}$$

Ici il faut faire confiance à $\alpha(n)$ qui doit croître très rapidement (d'où la chute de lissité de 42 utilisée). En multipliant les deux membres de E_4 par $\delta(n+1)$ on trouve

F (3) $$\delta(n+1)\, \|u_{n+1}\|_M \leq \frac{a^4\, \beta(n)\, \gamma(n-1)\, \delta(n+1)}{\alpha(n)} + a^2 \left(\delta(n)\|u_n\|_M\right)^2 \,.$$

Cette inégalité récurente montre que $\delta(n)\|u_n\|_M$ est le terme général d'une série rapidement convergente. Comme δ tend vers l'infini, il en est de même pour $\|u_n\|_M$.

On admirera ici la présence du terme quadratique $\|u_n\|_M^2$. (Grâce à l'emploi de la méthode de Newton.) Ce terme conduit à la condition F_1, avec le carré $(\delta(n))^2$ au deuxième membre, ce qui conduit à une fonction $\delta(n)$ croissante.

Si l'on renonce à la méthode de Newton, la condition F_1 devra être remplacée par

$$\beta(n)\,\delta(n+1) \leq \delta(n)$$

qui conduit (dès que $\beta(n) > 1$ à une fonction δ décroissante et si δ devrait tendre vers 0, la série $\delta(n)\|u_n\|_M$ pourrait être convergente, sans que $\|u_n\|_M$ le soit.

Nous laissons au lecteur interessé le soin de constater que dans les articles cités, on introduit un nombre N tel que les fonctions α, β, γ, δ puisse être de la forme

$$n \to N^{\lambda(3/2)^n}.$$

La fonction α correspond a un coefficient λ beaucoup plus grand que celui de β, γ, δ.

V. *L'exemple-pionnier*

Il s'agit de plonger isométriquement un tore T à deux dimensions muni d'une structure riemannienne, dans l'espace R^5 muni d'une structure euclidienne. Cela revient à définir une application u de T dans R^5, par 5 fonctions périodiques de deux variables

$$u_i\,(x_1, x_2) \qquad (i = 1, 2, \ldots n)\,.$$

Les coefficients du ds^2 sont trois fonctions $g_{\alpha\beta}(x_1,x_2)$ (avec α, β égaux à 1 ou 2) périodiques. Il s'agit de résoudre le système

$$\left(\frac{\partial U}{\partial x_\alpha}, \frac{\partial U}{\partial x_\beta}\right) = g_{\alpha\beta}$$

de 3 équations à 5 inconnues u_i. (Les parenthèses représentent un produit scalaire euclidien dans R^5.) Soit $\mathcal{E}^m$ l'espace des fonctions de classe C^m défini sur le tore. En posant:

$$\theta(U) = \left\{\left(\frac{\partial U}{\partial x_\alpha}, \frac{\partial U}{\partial x_\beta}\right) - g_{\alpha\beta}\right\}_{\substack{\alpha=1,2\\ \beta=1,2}}$$

On définit une application de classe C^{m-1} de $(\mathcal{E}^m)^5 = E$ dans $(\mathcal{E}^{m-1})^3 = \Phi$. La différentielle de θ est

$$D_U\theta[V] = \frac{\partial U}{\partial x_\alpha}, \frac{\partial V}{\partial x_\beta} + \frac{\partial U}{\partial x_\beta}, \frac{\partial V}{\partial x_\alpha} \qquad \alpha \text{ et } \beta = 1,2$$

n est évidemment pas injective (car $5 \neq 3$): mais Nash adjoint à l'équation linéarisée une condition de commodiré, $\left(\frac{\partial U}{\partial x_\alpha}, v\right) = 0$ qui lui fournit les deux équations qui lui manque.

Le problème posé par Nash consiste à l'examen de la possibilité du plongement isométrique, au voisinage d'un plongement régulier $\overset{\circ}{u}$. Alors

$$\det\left(\frac{\partial\overset{\circ}{u}}{\partial x_1}, \frac{\partial\overset{\circ}{u}}{\partial x_2}, \frac{\partial^2\overset{\circ}{u}}{\partial x_1^2}, \frac{\partial^2\overset{\circ}{u}}{\partial x_1\,\partial x_2}, \frac{\partial^2\overset{\circ}{u}}{\partial x_2^2}\right) \neq 0$$

et l'on peut construire un opérateur L, d'inversion qui permet de revenir de $(\mathcal{E}^{m-1})^3$ dans $(\mathcal{E}^{m-2})^5$.

On voit, en utilisant la méthode décrite que m doit être au moins égale à 44 pour être assurée que la réponse du problème sera au moins de classe C^2.

BIBLIOGRAPHIE

(*) BONIC. "Some Properties of Hilbert Scales" (en cours de publication).

NASH, J. "The imbedding problem for riemannian manifolds," *Ann. Math.* 63, 20-63 (1956).

[1] SCHWARTZ, J. T. *On Nash's implicit functional theorem,* Comm. Pure Math. 13, 509-530, (1960).

(*) [2] SCHWARTZ, J. T. *Non linear functional Analysis* (Courant Institute of Math-Sciences 1963-1964).

KOLMOGOROV, A. N. *Dokl. Akad. Nauk.* 98, 527 (1964).

Ce n'est qu'après le redaction de ce rapport que nous avons eu connaissance des articles marques (*).

KOLMOGOROV, A. N. General theory of dynamical systems ando-classical mechanics in *proceedings of the international Congress of Mathematics,* Amsterdam, 1954, ed. J. C. H. GERRETSEN and J. de GROOT (Amsterdam: Erven P. NOORDHOFF 1957) vol. 1, pp. 315-333.

ARNOLD, V. I. *Dokl. Akad. Nauk.* 137, 255-257 (1961).

Ibid 138, 13-15 (1961).

[1] MOSER, J. A new technique for the construction of solutions of non linear differential equations, *Proceedings of the Nat. Aca. of Sc.* vol. 47, n° 11, pp. 1824-1831, (1961).

(*) [2] MOSER, J. A rapidly convergent iteration method and non linear differential equations (I et II). *(Annali della Scuola Normale Superiore di Pisa*, vol. XX, 1966.

Ce n'est qu'après le redaction de ce rapport que nous avons eu connaissance des articles marques (*).

GENERALIZING THE HOPF-LEFSCHETZ FIXED POINT THEOREM FOR NON-COMPACT ANR-S

ANDRZEJ GRANAS

A continuous mapping $F: X \to Y$ between topological spaces X and Y is called *compact*, provided F maps X into a compact subset of Y. In this paper, we shall be concerned with the problem of the existence of fixed points for a compact map of an ANR for metric spaces into itself.*

Let X be an ANR for metric spaces. With a compact map $F: X \to X$ we shall associate an integer $\Lambda(F)$, defined in terms of the induced endomorphisms F_q of the homology of X and called the Lefschetz number of F.

In order to define the Lefschetz number $\Lambda(F)$, we shall invoke the theory of the trace, extended by J. Leray [9] for a class of endomorphisms of infinite-dimensional vector spaces. We shall not, however, state the definitions or properties needed in any more generality than will be necessary for our purposes.

The main result which we intend to present here is the following theorem: *if* $\Lambda(F) \neq 0$, *then the compact map* F *has a fixed point.*

This fact clearly implies several well-known fixed point theorems, both in functional analysis and topology, in particular, the Lefschetz Fixed Point Theorem for compact ANR-s and various forms of the Schauder Fixed Point Theorem.

---

* For a special kind of ANR-s (namely those which are r-dominated by subsets of Banach spaces) this problem was treated, in an implicit form, by F. Browder [3], in his paper concerned with fixed point theorems for Banach manifolds. The method used in that paper (based on the Mazur Lemma) could not be applied for arbitrary ANR-s.

The proof is based on the Hopf Fixed Point Theorem for polyhedra [6] and the theorems of Kuratowski [7] and Wojdysławski [11] concerning the embedding of a metric space into a normed space.

§1. *The trace.* In what follows, we shall consider vector spaces only over the field Q of rational numbers.

Given an endomorphism $\phi: E \to E$ of a finite-dimensional vector space E, we denote by $tr(\phi)$ the trace of the endomorphism ϕ.

We recall the following well-known property of the trace:

(A) *assume that we are given finite-dimensional vector spaces* E', E'' *and a commutative diagram*

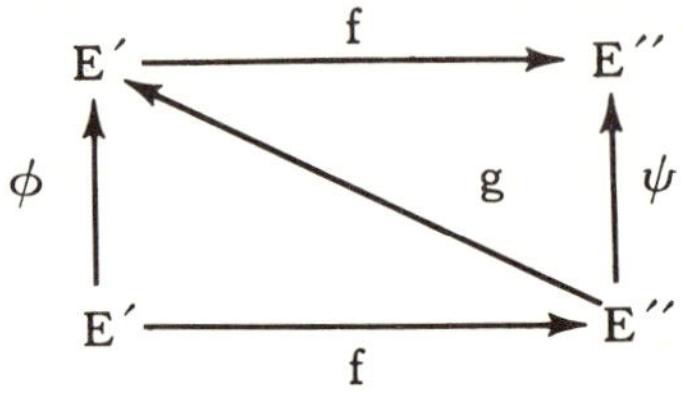

of linear maps. Then $tr(\phi) = tr(\psi)$.

A graded vector space $E = \{E_q\}_{q=0}^{\infty}$ is said to be of a *finite type,* provided all E_q are of finite dimension and $E_q = 0$ for almost all q. Let E be of a finite type and $\phi = \{\phi_q\}: E \to E$ be an endomorphism of degree 0. We let

$$tr(\phi) = \{tr(\phi_q)\}$$

$$\lambda(\phi) = \sum_{q=0} (-1)^q \, tr(\phi_q)$$

and call $\lambda(\phi)$ the Lefschetz number of ϕ.

Let E be an arbitrary vector space. Call an endomorphism $\phi: E \to E$ *finite-dimensional,* provided $\dim(\mathrm{Im}\,\phi) < +\infty$. For a finite-dimensional ϕ, let E' be a finite-dimensional subspace of E containing $\mathrm{Im}\,\phi$ and $\phi': E \to E'$ be the contraction* of ϕ to the pair (E', E').

* Let $\mathcal{C}$ be the concrete category and let $f: X \to Y$ be a map in $\mathcal{C}$ such that $f(A) \subset B$, where $A \subset X$ and $B \subset Y$. By the *contraction* of f to the pair (A, B) we shall understand a map $f^*: A \to B$ in $\mathcal{C}$ defined by $f^*(a) = f(a)$, for $a \in A$. A contraction of f to the pair (A, Y) is simply the restriction $f|A$ of f to A.

We then define the (generalized) *trace* $\mathrm{Tr}(\phi)$ of ϕ by putting

$$\mathrm{Tr}(\phi) = \mathrm{tr}(\phi') .$$

It follows clearly from (A) that $\mathrm{Tr}(\phi)$ does not depend on the choice of the space E'.

Let $E = \{E_q\}$ be a graded vector space and

$$\phi = \{\phi_q\}: E \to E$$

be an endomorphism of degree 0. We shall say that ϕ is a *Lefschetz endomorphism* or L-endomorphism, provided the graded vector space $\mathrm{Im}\,\phi = \{\mathrm{Im}\,\phi_q\}$ is of a finite type.

In this case we define the (generalized) *Lefschetz number* $\Lambda(\phi)$ of ϕ by putting

$$\Lambda(\phi) = \sum_{q=0} (-1)^q \,\mathrm{Tr}(\phi_q) .$$

For notational convenience we let $\mathrm{Tr}(\phi) = \{\mathrm{Tr}(\phi_q)\}$. Clearly, if E is of a finite type and $\phi: E \to E$, then $\mathrm{tr}(\phi) = \mathrm{Tr}(\phi)$ and $\lambda(\phi) = \Lambda(\phi)$.

LEMMA 1. *Let $E' = \{E'_q\}$ be a graded vector space of a finite type and assume that we are given a commutative diagram of graded vector spaces and linear maps*

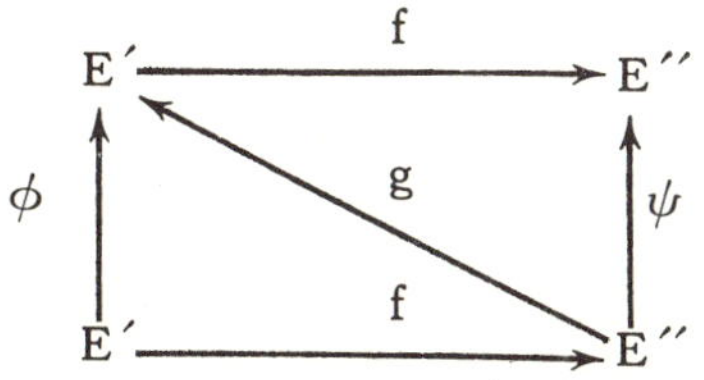

Then, ψ is a Lefschetz endomorphism, and we have in this case $\Lambda(\phi) = \Lambda(\psi)$.

Proof: From the commutativity of the above diagram, we have $\mathrm{Im}\,\psi \subset \mathrm{Im}\,f$. Since E' is of a finite type, this implies that ψ is a Lefschetz endomorphism. Consequently, we have the commutative diagram

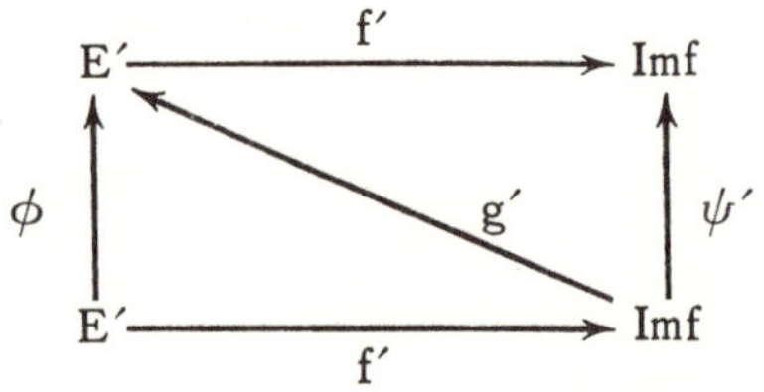

in which f', g', ψ' stand for the obvious contractions. In view of the property (A), we have $\mathrm{tr}(\phi) = \mathrm{tr}(\psi') = \mathrm{Tr}(\psi)$. Consequently, $\Lambda(\phi) = \Lambda(\psi)$ and the proof is completed.

§2. *The Lefschetz maps.* In what follows, we shall denote by H the singular homology functor with rational coefficients from the category of topological spaces and continuous maps to the category of graded vector spaces and linear maps of degree 0. Thus, $H(X) = \{H_q(X)\}$ is a graded vector space, $H_q(X)$ being the q-dimensional singular homology group of a space X. For a map $f: X \to Y$, $H(f)$ is the induced linear map $f_* = \{f_q\}: H(X) \to H(Y)$, where $f_q: H_q(X) \to H_q(Y)$.

A map $f: X \to X$ will be called a *Lefschetz map* or L-map provided $f_*: H(X) \to H(X)$ is a Lefschetz endomorphism. In this case, we define the Lefschetz number $\Lambda(f)$ of f by putting

$$\Lambda(f) = \Lambda(f_*) .$$

A topological space X will be called of a *finite type* provided the graded vector space $H(X)$ is of a finite type.

LEMMA 2. *Let* P *be a space of a finite type and assume that we are given a commutative diagram of spaces and maps*

$\mathfrak{D} =$

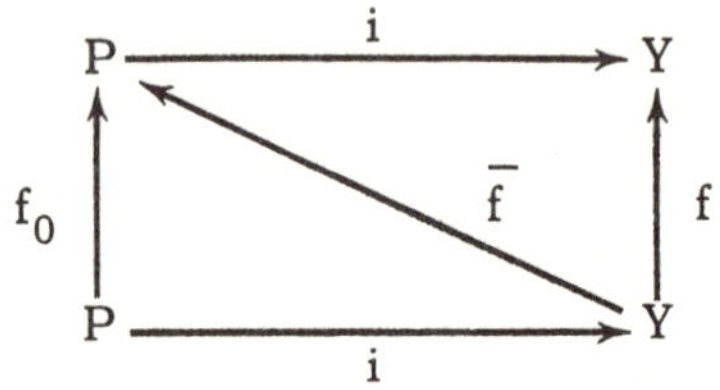

in which $i:P \to Y$ *stands for the inclusion. Then* f *is a Lefschetz map and we have in this case* $\Lambda(f) = \Lambda(f_0)$.

Proof. Applying the homology functor to the diagram $\mathfrak{D}$, we obtain the commutative diagram

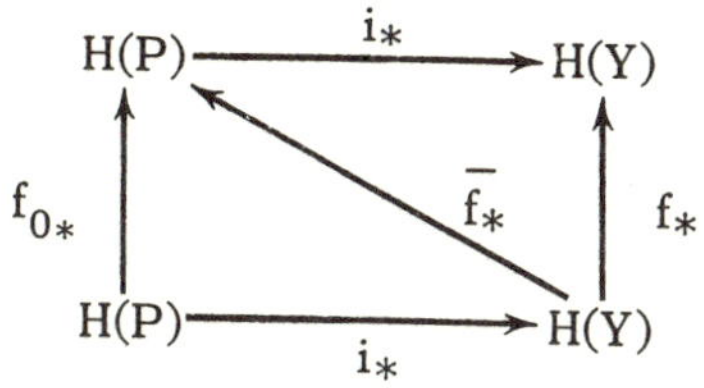

which satisfies the assumption of Lemma 1.

Thus, $\Lambda(f_{0*}) = \Lambda(f_*)$ and, consequently, $\Lambda(f_0) = \Lambda(f)$. The proof is completed.

§3. r-*maps and* s-*maps*. Let $\mathcal{C}$ be a category and denote by $\mathcal{C}(X)$ the set of maps $X \xrightarrow{f} X$ in $\mathcal{C}$. Assume that we are given a pair of maps $s:X \to Y$ and $r:Y \to X$ in $\mathcal{C}$. Following Borsuk [1], we say that r is an r-*map for* s, and s an s-*map for* r provided $rs = 1_X$. A map $r:Y \to X$ (resp. $s:X \to Y$) for which there exists an s-map (resp. an r-map) is called simply an r-map (resp. an s-map).

Let $s:X \to Y$ be an s-map and $r:Y \to X$ an r-map for s. We define the maps between the sets

$$\sigma: \mathcal{C}(X) \to \mathcal{C}(Y)$$

$$\rho: \mathcal{C}(Y) \to \mathcal{C}(X)$$

by the following formulas

(1) $$\rho(f) = rfs \qquad \text{for } f \in \mathcal{C}(Y)$$

(2) $$\sigma(f_0) = sf_0r \qquad \text{for } f_0 \in \mathcal{C}(X)\,.$$

LEMMA 3. *Putting* $f = \sigma(f_0)$ *for each* $f_0 \in \mathcal{C}(X)$, *we establish* 1-1 *correspondence* $f_0 \longleftrightarrow f$ *between the sets* $\mathcal{C}(X)$ *and* $\sigma(\mathcal{C}(X)) \subset \mathcal{C}(Y)$.

If $\mathcal{C}$ *is the category of graded vector spaces, then*

1° f_0 *is an* L-*endomorphism* $\Longleftrightarrow$ f *is an* L-*endomorphism*

2° *if* f_0 *is an* L-*endomorphism, then* $\Lambda(f_0) = \Lambda(f)$.

If $\mathcal{C}$ *is the category of topological spaces, then*

3° f_0 *is compact* $\Longleftrightarrow$ f *is compact*

4° f_0 *has a fixed point* $\Longleftrightarrow$ f *has a fixed point*

5° f_0 *is an* L-*map* $\Longleftrightarrow$ f *is an* L-*map*

6° *if* f_0 *is an* L-*map, then* $\Lambda(f_0) = \Lambda(f)$.

Proof. The first assertion follows from the fact that ρ is an r-map for σ in the category of sets. 1° and 3° are evident. The proof of 2° is similar to that of Lemma 1.

To prove 4°, note that if x is a fixed point for f_0, then so is s(x) for f; if y is a fixed point for f, then so is r(y) for f_0.

5° and 6° follow from 1° and 2°, the same way as Lemma 2 follows from Lemma 1.

§4. *Compact maps into the special* ANR-s. In this and in the next two sections by a space we shall understand a metrizable space.

A space X is called an ANR (respectively AR) provided that for each embedding $h:X \to Y$ such that h(X) is closed in Y, there exists a retraction $r:U \to h(X)$, of an open set $U \subset Y$ onto h(X) (respectively a retraction $r:Y \to h(X)$).

A space Y will be called a *special* ANR provided Y is an open subset of a convex set Z contained in a normed space.*)

A mapping $F:X \to Y$ into a special ANR will be called *finite dimensional* provided the image F(X) is contained in a finite polyhedron $P \subset Y$. It turns out that every compact mapping into a special ANR can be uniformly approximated by finite dimensional mappings. More precisely, we have the following

*) The fact that every special ANR is an ANR will not be used in our discussion.

THEOREM 1 *(Approximation Theorem). Let* Y *be a special* ANR *and* $F:X \to Y$ *be a compact mapping of a space* X *into* Y. *Then, for each* $\varepsilon > 0$, *there exists a finite dimensional mapping* $G:X \to Y$ *such that*

(i) *for all* $x \in X$ *we have*

$$\|F(x) - G(x)\| < \varepsilon$$

(ii) G *is homotopic to* F.

Proof. Let $\varepsilon > 0$ be given. Let us put now $Y_0 = \overline{F(X)} \subset Y$. Since Y_0 is compact (by the assumption of the theorem) and Y is open in a convex set Z, there exists a constant δ satisfying

$$0 < \delta < \varepsilon \tag{1}$$

and such that each ball

$$V(y_0, 2\delta) = \{y \in Z, \|y-y_0\| < 2\delta\} \tag{2}$$

with center $y_0 \in Y_0$ and radius 2δ is entirely contained in Y.

Now take an arbitrary $\frac{\delta}{2}$ –net $N = \{y_1, y_2, \ldots, y_k\}$ of the compact set Y_0 and put for each $i = 1, 2, \ldots, k$

$$V_i = V(y_i, \frac{\delta}{2}), \quad V_i' = V(y_i, 2\delta) .$$

Evidently, all the balls V_i and V_i' are convex subsets of Y and $V = \bigcup_{i=1}^{k} V_i$ covers Y_0.

Define the partition of unity $\{\lambda_i\} \bigcup_{i=1}^{k}$ on V by putting

$$\mu_i(y) = \max \{0, \frac{\delta}{2} - \|y-y_i\|\} \tag{3}$$

$$\lambda_i(y) = \frac{\mu_i(y)}{\sum_{j=1}^{k} \mu_j(y)} \text{ for } y \in V \tag{4}$$

Let us put for each $x \in X$

$$G(x) = \sum_{j=1}^{k} \lambda_i(F(x))\, y_i \tag{5}$$

It follows from (1) and the definition of G that

(6) $$\|G(x) - F(x)\| < \varepsilon \text{ for all } x \in X$$

and that the values of G are in a simplicial complex P with the vertices $y_1, y_2, \ldots, y_k$.

It follows from (5) that the points y_i, which appear in a convex combination (5) for some x, belong to one of the balls V_i'. Since all V_i' are convex, we conclude that $P \subset Y$.

To prove the assertion (ii), note that for each $x \in X$, the points $G(x)$ and $F(x)$ belong to one of the balls V_i', say to V_j'. Since V_j' is convex subset of Y, the family of maps g_t defined by the formula

$$g_t(x) = tG(x) + (1-t)\,F(x) \text{ for } (x \cdot t) \in X \times I$$

has values in Y. Thus $g_t : X \to Y$ is a homotopy joining G with F and the proof is completed.

§5. *The main theorem.* In what follows we shall make use of the following elementary fact:

LEMMA 4. *Let* $F:Y \to Y$ *be a compact map of a metric space* Y *into itself and assume that* F *is a uniform limit of a sequence* $\{F_n\}$ *of maps* $F_n:Y \to Y$. *If each* F_n *has a fixed point, then so does the map* F.

Proof. Let $\{y_n\}$ be a sequence of points such that $F_n(y_n) = y_n$. From the assumption it follows that for almost all n we have

(1) $$\rho(F(y_n), y_n) < \frac{1}{n} .$$

Since F is compact, we may assume, without loss of generality, that

(2) $$\lim_{n=\infty} F(y_n) = y \in Y .$$

It follows from (1) and (2) that $\lim_{n=\infty} y_n = y$ and hence, by continuity of F, we have

$$\lim_{n=\infty} F(y_n) = F(y) .$$

Comparing (2) and (3) we obtain $F(y) = y$ and the proof is completed.

THEOREM 2. *Let* Y *be a special* ANR *and* $F: Y \to Y$ *be a compact map. Then* (i) F *is a Lefschetz map,* (ii) $\Lambda(F) \neq 0$ *implies that* F *has a fixed point.*

Proof. Theorem 1 implies that F is a uniform limit of a sequence $\{F_n\}$ of mappings $F_n: Y \to Y$ such that

(4) $$F \sim F_n \text{ for all } n$$

(5) $$F_n(Y) \subset P_n ,$$

where P_n is a finite polyhedron.

Denote by $\overline{F}_n: P_n \to P_n$ the obvious contraction of F_n. Lemma 2 implies that each F_n is a Lefschetz map, and hence, in view of (4), so is the map F. This completes the proof of (i).

To prove (ii), assume that $\Lambda(F) \neq 0$. In view of (4), we have $\lambda(F_n) \neq 0$. Lemma 2 implies that $\Lambda(F_n) = \lambda(\overline{F}_n) \neq 0$. This implies, in view of the Hopf Fixed Point Theorem [6], that for every n there exists a point $y_n \in P_n$ such that

$$y_n = \overline{F}_n(y_n) = F_n(y_n) .$$

It follows now from Lemma 4 that F has a fixed point and the proof of the theorem is completed.

Now we state the main result of the paper:

THEOREM 3. *Let* X *be an arbitrary* ANR *and* $F_0: X \to X$ *be a compact map. Then* (i) F_0 *is a Lefschetz map,* (ii) $\Lambda(F_0) \neq 0$ *implies that* F_0 *has a fixed point.*

Proof. In view of the theorems of Kuratowski [7] and Wojdysławski [11], X is r-dominated by a special ANR Y. Using the notation of the Section 3, let us put $F = \sigma(F_0)$. By Lemma 3, $F: Y \to Y$ is compact and, hence, by Theorem 2, is a Lefschetz map. Applying Lemma 3 again, we conclude that F_0 is a Lefschetz map. Thus (i) is proved.

To prove (ii), assume that $\Lambda(F_0) \neq 0$. By Lemma 3, $\Lambda(F_0) = \Lambda(F) \neq 0$. This, in view of Theorem 2, implies that F has a fixed point. Applying

Lemma 3 for the last time, we conclude that F_0 has a fixed point, and the proof of the theorem is completed.

§6. *Corollaries.* We list now a few immediate consequences of the main theorem:

COROLLARY 1 (*The Lefschetz Fixed Point Theorem* [8]). *Let* X *be a compact* ANR *and* $f:X \to X$ *be a map. If* $\lambda(f) \neq 0$, *then* f *has a fixed point.*

A space X is said to have the *fixed point property in the narrow sense* (cf. [5]), provided every compact map $F : X \to X$ has a fixed point.

COROLLARY 2. *Acyclic* ANR-s *and in particular* AR-s *have the fixed point property in the narrow sense.*

COROLLARY 3 (*The Schauder Fixed Point Theorem* [10]). *Let* X *be a convex (not necessarily closed) subset of a normed space (or of a metrizable locally convex space). Then* X *has the fixed point property in the narrow sense.*

Proof. By the Theorem of Dugundji [4], X is an AR. Our assertion follows therefore from the Corollary 2.

COROLLARY 4 (*The Birkhoff-Kellogg Theorem* [2]). *Let* S *be the unit sphere in an infinite dimensional normed space* E *and* $F:S \to E$ *be a compact map such that*

(1) $$\|F(x)\| \geq a > 0 \text{ for all } x \in S .$$

Then there is a positive λ, *such that* $F(x) = \lambda x$ *for some* $x \in S$.

Proof. Define $\bar{F}:S \to S$ by

$$\bar{F}(x) = \frac{F(x)}{\|F(x)\|} .$$

Because of (1), $\bar{F}$ is compact. Since the unit sphere S is an AR, by a theorem of Dugundji [4], there is a point $x \in S$ such that

$$x = \bar{F}(x) = \frac{F(x)}{\|F(x)\|} .$$

Thus, $F(x) = \lambda x$, with $\lambda = \|F(x)\|$, and the proof is completed.

§7. *Further remarks.* The main theorem given in the previous section can be generalized to non-metrizable case.

Call a topological space Y a *special Borsuk space* provided Y is an open set in a convex set Z lying in a locally convex linear topological space E. A topological space X is said to be a *Borsuk space* provided X is r-dominated by a special Borsuk space Y.

We have the following theorem:

THEOREM 4. *Let* $F: X \to X$ *be a compact map of a Borsuk space* X *into itself. Then* F *is a Lefschetz map and* $\Lambda(F) \neq 0$ *implies that* F *has a fixed point.*

A compact topological space X is called an ANR *for normal spaces* provided for each embedding $h: X \to Y$ into a normal space Y the set $h(Y)$ is a neighbourhood retract of Y. Every compact ANR for normal spaces (being homeomorphic with a neighbourhood retract of a Tychonoff cube) is evidently a Borsuk space. This implies, in view of Theorem 4, the following generalization of the Lefschetz fixed point theorem for non-metrizable compact ANR-s:

COROLLARY 1. *Let* X *be a compact* ANR *for normal spaces and* $f: X \to X$ *be a map. Then* f *is a Lefschetz map and* $\Lambda(f) \neq 0$ *implies that* f *has a fixed point.*

As a consequence of Theorem 4, we note also the following generalization of the Tychonoff fixed point theorem:

COROLLARY 2. *Let* X *be a convex (not necessarily closed) subset of a locally convex linear topological space. Then* X *has the fixed point property in the narrow sense.*

BIBLIOGRAPHY

[1] K. Borsuk, *On the topology of retracts,* Annals of Math. 48 (1947), pp. 1082-1094.

[2] D. Birkhoff and O. D. Kellogg, *Invariant points in function spaces,* Trans. Amer. Math. Soc. 23 (1922), pp. 96-115.

[3] F. Browder, *Fixed Point Theorems for Infinite Dimensional Manifolds*, Trans. Amer. Math. Soc. (1965), pp. 179-193.

[4] J. Dugundji, *An Extension of Tietze's Theorem*, Pacific Journ. Math. 1 (1951), pp. 353-367.

[5] A. Granas, *The theory of compact vector fields and some of its applications to topology of functional spaces*, Rozprawy Matematyczne XXX, Warszawa 1962.

[6] H. Hopf, *Eine Verallgemeinerung der Euler-Poincareschen Formel*, Nachr. Ges. Wiss. Gotingen (1928), pp. 127-136.

[7] K. Kuratowski, *Quelques problèmes concernant les espaces métriques non-séparables*, Fund. Math. 25 (1935), pp. 534-545.

[8] S. Lefschetz, *On locally connected and related sets*, Annals of Math. 35 (1934), pp. 118-129.

[9] J. Leray, *Théorie des points fixes: Indice total et nombre de Lefschetz*, Bull, Soc. Math. France 87 (1959), pp. 221-233.

[10] J. Schauder, *Der Fixpunktsatz in Funktionalräumen*, Studia Math, 2 (1930), pp. 171-180.

[11] M. Wojdysławski, *Retractes absolus et hyperespaces des continus*, Fund. Math. 32 (1939), pp. 184-192.

SOME QUESTIONS IN THE DIMENSION THEORY OF INFINITE DIMENSIONAL SPACES

by

DAVID W. HENDERSON

In this paper, 'space' means 'compact metric space'. The (large inductive) dimension, Ind(X), of a space, X, is defined by transfinite induction as follows: (a) Ind(X) = −1, if X is the empty set. (b) For each ordinal α, Ind(X) $\leq \alpha$, if each closed subset of X has arbitrarily small neighborhoods whose boundaries have dimension $< \alpha$. (c) Ind(X) = α, if Ind(X) $\leq \alpha$ and if, for each $\beta < \alpha$, it is not true that Ind(X) $\leq \beta$. (d) If Ind(X) either fails to exist or is infinite, then X is said to be infinite-dimensional.

Question 1. *Does every infinite-dimensional space contain subsets of each finite dimension'* [Clearly each non-empty space contains a compact 0-dimensional subset. However, [37] and [43] describe infinite-dimensional spaces which contain no compact α-dimensional subsets, for $\alpha > 0$. See also [41].]

A map, $f: X \to I^n$, from a space to an n-cell is said to be *essential*, if $f \mid f^{-1}(\mathrm{Bd}\, I^n)$ can not be extended to a map taking all of X onto $\mathrm{Bd}\, I^n$. A map, $f: X \to I^\omega$, of a space to the Hilbert Cube is said to be *essential*, if $f \mid f^{-1}(F)$ is essential for each finite-dimensional face, F, of I^ω. If X has an essential mapping onto the Hilbert Cube then X is said to be *strongly infinite-dimensional.*

A space is *countable-dimensional* if it is the union of a countable number of 0-dimensional subsets.

Question 2. (Alexandroff) *Is each non-countable-dimensional space strongly infinite-dimensional?* [It is known (see [4, pp. 48-51] and [36, Chapter III, pp. 18-20]) that no countable-dimensional (compact) space is strongly infinite-dimensional and that each (compact) space for which Ind(X) exists is countable-dimensional. It is conjectured in [43] that the space constructed in [43] answers Question 2 in the negative.]

Question 3. *Which spaces,* X, *are homeomorphic to sets that separate some open subset of the Hilbert Cube?* [It can be shown that X cannot be countable-dimensional. In addition, X must contain closed n-dimensional subsets for each finite n, for if X separates the Hilbert Cube then it must separate some (n+1)-dimensional cube in the Hilbert Cube and the intersection of X with this (n+1)-cube would have dimension n or n+1. ([42] describes strongly infinite-dimensional spaces without n-dimensional closed subsets.)]

CORNELL UNIVERSITY

BIBLIOGRAPHY OF THE DIMENSION THEORY OF INFINITE-DIMENSIONAL SPACES

(The square brackets at the end of some of the references refer to the review of the article in *Mathematical Reviews (MR).)*

1. Alexandroff, P. S., Dimensionstheorie, *Math. Ann., 106* (1932), 161-238.
2. Hurewicz, W., Une remarque sur l'hypotheses du continie, *Fund. Math., 19* (1932), 8-9.
3. Freudenthal, H., Entwicklungen von Raumen and ihren Gruppen, *Compositio Mathematica, 4* (1937), 10-234.
4. Hurewicz and Wallman, *Dimension Theory,* Princeton University Press, Princeton, N. J., 1941.
5. Van Heemert, A., The existence of 1- and 2-dimensional subspaces of a compact metric space, *Amsterdam Acad., van Wetenschapen-Indagationes Mathematicae, 8* (1946), 564-569. (This paper is incorrect.)
6. Toulmin, G. H., Shuffling ordinals and transfinite dimension, *Proc. London Math. Soc., 4* (1954), 177-195.
7. Alexandrov, P. S., The present status of the theory of dimension, *Amer. Math. Soc., Translations,* Series 2, vol. *1* (1955), 1-26.
8. Tumarkin, L. A., On infinite-dimensional Cantor manifolds (Russian), *Dokl. Akad. Nauk SSSR, 115* (1957), 244-246. This is superceded by [36].)
9. Nagata, J., On the countable-dimensional spaces, *Proc. Japan Acad., 34* (1958), 146-149.
10. Levsenko, B. T., Sklyarenko, E. G., Smirnov, Ju. M., Concerning infinite-dimensional spaces (Russian); *Uspekhi matem. Nauk 13,* No. 5 (83), 195.
11. Levsenko, B. T., Strongly infinite-dimensional spaces (Russian), *Vestnik Moskov Univ. Ser. Mat. Astr. Fiz. Nim.* (1959), No. 5, 219 [*MR 22* (4053)].

12. Nagami, Keio, Finite-to-one closed mappings and dimension II, *Proc. Japan Acad.*, *35* (1959), 437-439 [*MR 22* (4025)(1961)].

13. Sklyarenko, E. G., On dimensional properties of infinite-dimensional spaces, *Izv. Akad. Nauk, SSSR Ser. Mat. 23* (1959), 197 [*MR 21* (5179) (translation: *AMS Transl.* (2) *21* (1962), 35-50).

14. Sklyarenko, E. G., Universal spaces for certain classes of ∞-dim spaces, *Izd. Akad. Nauk SSSR, Sem. Mat. 23* (1959), 185 [*MR* (5178)], (translated in *AMS Translations* (2) *21* (1962), 21-33.

15. Mardesic, Sibe, Covering dimension and inverse limits of compact spaces, *Illinois J. of Math.*, *4* (1960), 278-291.

16. Nagata, J., On the countable sum of 0-dimensional metric spaces, *Fund. Math.*, *48* (1960), 1-14 [*MR 22*(5028)].

17. Nagata, J., Two remarks on dimension theory for metric spaces, *Proc. Japan Acad.*, *36* (1960), 53 [*MR 22* (9963)].

18. Nagami, Keto, Mappings of finite order and dimension theory, *Japan J. Math.*, *30* (1960), 25-54 [*MR 25* (5494)]. (This paper contains a general development of dimension theory for metric spaces.)

19. Sersnev, M., Strong dimension of mappings and the associative characterization of dimension for arbitrary metric spaces, *Soviet Math. Dokl.*, *1* (1960), 1267-1269 [*MR 23* (A2865)]. (The details in Mat. Sb. (N. S.) *60* (102) (1963), 207-218 [*MR 28* (585)].

20. Tumarkin, L. A., On the decomposition of spaces into a countable number of 0-dimensional sets (Russian), *Vestnik Moskov. Uhiv. Sec. I Mat. Mek.* 1960, No. 1, 25-32 [*MR 25* (554)].

21. Alexandroff, P. S., Results in Topology in the last 25 years, *Russian Math. Surveys, 15* (1960) (translated from Uspekhi matem. Nauk 15 (1960), 25-97).

22. Sklyarenko, E. G., Representation of infinite-dimensional compacts as an inverse limit of polyhedra, *Soviet Math. Dokl.*, *1* (1960), 1147-49, 134, (1960), 773-775].

23. Alexandroff, P. S., On some results concerning topological spores and their continuous mappings, General Topology and its Relations to Modern Algebra and Analysis (*Proc. Sympos., Prague,* 1961). pp. 41-54. Academic Press, New York; [*MR 26* (3003)].

24. Arhangel'skii, A., The ranks of systems of sets and the dimensions of spaces, *Soviet Math. Dokl.,3* (1962), 456-459 [*MR 24* (A2368)]. (The details in *Fund. Math., 52* (1963), 257-275.)

25. Levsenko, B. T., On ∞-dimensional spaces, *Soviet Math. Dokl., 2* (1961), 915, [*MR 24* (A 1708)].

26. Sklyarenko, E. G., Some remarks on spaces having an infinite number of dimensions (Russian), *Dokl, Akad. Nauk. SSSR 126* (1959), 1203 [*MR 22* (1885)(1961)].

27. Smirnov, Ju. M., Some remarks on transfinite dimension, *Dokl. Akad. Nauk. SSSR, 141* (1961), 814 [*MR 26* (6934)]. (Translated in *Soviet Math. Dokl., 2* (1961), 1572.)

28. Smirnov, Ju. M., On dimensional properties of infinite-dimensional spaces, 334-336. General Topologyand its Relations to Modern Algebra and Analysis (*Proc. Symp., Prague,* 1961).

29. Smirnov and Sklyarenko, Some questions in dimension theory (Russian), *Proc. 4th All-Union Math. Congress* (Leningrad, 1961), vol. *I*, pp. 219-226.

30. Tumarkin, L. A., Concerning infinite-dimensional spaces, General Topology and its Relations to Modern Algebra and Analysis (*Proc. Symp., Prague,* 1961), pp. 352-353.

31. Sklyarenko, E. G., Homogeneous spaces of an infinite number of dimensions, *Dokl. Akad. Nauk. SSSR,* (AHCCCP), *141* (1961), 811-813. (Translated in *Soviet Math. Dokl. 2* (1961), 1569.)

32. Pasynkov, B., On the spectra and dimensionality of topological spaces, *Mat. Sb.* (N.S.) *57* (99)(1962), 449-276. [*MR 26* (1856)].

33. Sklyarenko, E. G., Two theorems on infinite-dimensional spaces, *Soviet Math. Dokl., 3* (1962), 547-550 [*MR 25* (A 2367)].

34. Smirnov, Ju. M., On transfinite dimension (Russian), *Mat. Sb.* (N.S.) *58* (100) (1962), 415-522.

35. Tumarkin, L. A., On strongly and weakly infinite-dimensional spaces (Russian), *Vestnik, Moskov Univ. Ser I, Mat. Meh.* (1963) No. 5, 24-27 [*MR 27* (5239)].

36. Alexandroff, P. S., On some basic directions in general topology, *Russian Math. Surveys, 19* (1964), No. 6, 1-39. (Translated from *Uspehki Nauk, 19* (1964) No. 6.)

37. Henderson, D. W., An infinite-dimensional compactum with no positive-dimensional compact subsets, (multility) Institute for Advanced Study, 1965.

38. Henderson, D. W., Every compactum contains a connected 1-dimensional subset, Abstract 622-69, *Notices of AMS, 12* (1965), 347.*

39. Nagata, J., *Modern Dimension Theory,* Interscience, New York, 1965.

40. Levsenko, B. T., Spaces with a transfinite number of dimensions (Russian), *Mat. Sb.* (N.S.) *67* (109) (1965), 255-266 [*MR 31* (6213)]. (6213)].

41. Henderson, D. W., Finite-dimensional subsets of infinite-dimensional spaces, Topology Seminar, Wisconsin, 1965, *Annals of Math Studies 60,* Princeton, N. J., 1966.

42. Henderson, D. W., Each strongly infinite-dimensional space contains a hereditarily infinite-dimensional subset, *Amer. J. Math., 89* (1967).

43. Henderson, D. W., An infinite-dimensional compactum with no positive-dimensional compact subsets—a simpler construction, *Amer. J. Math., 89* (1967).

44. Henderson, D. W., A lower bound for transfinite dimension, (submitted).

45. Henderson, D. W., A new transfinite dimension, (submitted).

*The proof claimed in this abstract is faulty

ON THE CONTINUITY OF
BEST APPROXIMATION OPERATORS [1]

by

R. B. HOLMES

Let X be a Banach space and M a Chebyshev subspace of X. The best approximation operator (metric projection, nearest point map) supported by M is denoted P_M. It is coming to be realized [6, 16] that considerable information about the "smoothness" of P_M is reflected in the structure of $\ker P_M = \{x \in X: P_M(x) = \theta\}$. For example, P_M is linear iff $\ker P_M$ is convex [16]. Other results on the relation between P_M and its kernel in the cases that P_M is Lipschitzian and/or differentiable may also be found in [16]. The object of the present paper is to study the relation between the continuity of best approximation operators (and related maps) and the topological structure of $\ker P_M$ (and related manifolds).

It is familiar that P_M is generally not linear when X is not a Hilbert space. Quite recently it has been shown that P_M may even be discontinuous. Examples have been given by Holmes and Kripke (for X a rotund isomorph of ℓ^∞) [16] and Wulbert (for $X = \ell^1$)—see [6].[2] In both examples, codim M = 2 while X is non-reflexive. We know of no example of a discontinuous P_M which acts on a reflexive space; however, some of the

[1] Presented in part to the American Mathematical Society, Feb. 9, 1967, Abs. 67-313, Not. AMS, 14 (1967), 422.

[2] The first example of this behavior is implicitly due to Lindenstrauss [24, Ch. VII]. In an unpublished manuscript P. Morris has established (among other things) the following result: let K be a compact Hausdorff space containing infinitely many points and suppose that C(K) contains a Chebyshev subspace M of finite codimension greater than unity. Then P_M is discontinuous.

results of §2 of this paper tend to suggest that such an example may exist.

Suppose now that X is both reflexive and rotund, so that P_M is always well defined. Then X^* is smooth and hence the norm of X^* has at every non-zero point a Gateaux differential. Let us assume that this dual norm is actually Fréchet differentiable. Then it is a consequence of a theorem of Fan and Glicksberg [11] and some well-known duality results that every P_M is continuous on X. Now the dual norm is Fréchet differentiable iff the spherical image map (Cudia [8]) T is continuous from X^* to $X^{**} = X$. Thus the continuity of T on X^* implies the continuity of (all) P_M on X. It is not known whether the converse of this statement is generally valid. We have, however, the following result in this direction: *if* X *is rotund and reflexive, then* P_M *is continuous for all finite codimensional* $M \subset X$ *iff* T *is continuous in the finite topology on* X^*. The proof uses some results from the theory of monotone operators [2, 3, 20]. This result and others similar are developed in §2 below wherein we study relationships between the continuity of P_M and structural properties of the quotient space, the annihilator, and the dual space.

In the first section we study P_M and its kernel from a topological viewpoint. We obtain some new "weak linearity" properties of P_M (cf. [16]), namely that P_M is *open* if it is continuous and that it is (sometimes at least) *weakly* continuous. The main result of the section is the following characterization theorem, which proves to be quite useful: P_M *is continuous iff* ker P_M *is homeomorphic to* X/M *under the quotient map.*

Some conjectures and statements of open problems occur throughout the paper.

Finally, we want to emphasize that the basic assumptions which we have made about our set of approximators M are fairly restrictive. That is, we have assumed both that M is a Chebyshev set *and* a linear subspace. On the one hand, this is a situation of frequent occurrence and results obtained often possess an artistic appeal not always found in general. As only one particular illustration of this, we recall that there is a very close and natural connection between the uniqueness and continuity of best approximation from subspaces and the uniqueness and continuity of Hahn-Banach extensions from the corresponding annihilators (at least for reflexive spaces—see [29, §1] and [24, Ch. VII]). On the other hand, there is certainly considerable interest in approximation situations in which one or both of our basic hypotheses is lacking. For results on approximation out of general Chebyshev sets, we may refer to [10, 22, 35, 39], and for non-unique best approximation to [1, 33], and the literature therein cited.

§1. *The Best Approximation Operator and its Kernel.*

Throughout this section X will be a real Banach space, X* its dual space of continuous linear functionals on X, M a (closed) linear Chebyshev subspace of X, and X/M its quotient space. The best approximation operator (BAO) supported by M will always be denoted by P_M; it is defined by $P_M(x) = m^* \epsilon M$ if $\|x - m^*\| < \|x - m\|$ for every $m \epsilon M$, $m \neq m^*$. It is immediate that P_M is a closed idempotent mapping of X onto M. Also P_M is clearly homogeneous ($P_M(cx) = cP_M(x)$ for all scalars c and $x \epsilon X$), bounded ($\|P_M(x)\| \leqq \|P_M(x) - x\| + \|x\| \leqq 2\|x\|$), and additive (mod M) ($P_M(x+y) = P_M(x) + P_M(y)$ if x and y belong to the same coset in X/M). In particular, P_M is continuous whenever it is additive. On the other hand if P_M is *additive* for *all* M then X must be a Hilbert space [28], and, of course, conversely. This result has been sharpened by Rudin-Smith [31], Holmes [15], and Hirshfeld [14].

In spite of the fact that P_M is generally non-linear we still have an important "direct sum" decomposition of X induced by P_M. Namely, every $x \epsilon X$ can be written *uniquely* as $x = m + k$ where $m \epsilon M$, $k \epsilon \ker P_M$. For clearly we can write $x = P_M(x) + (x - P_M(x))$. Thus the uniqueness part is reduced to showing that if $k_1, k_2 \epsilon \ker P_M$ and $k_1 - k_2 \epsilon M$, then $k_1 = k_2$, or in other words, that $(\ker P_M + \ker P_M) \cap M = \{\theta\}$, where θ is the zero element of X. In this form the result is due to Singer and Ptak (see [5, §2] for references and a proof), and actually characterizes the Chebyshev property of M.

Suppose now that P_M is continuous. Then the direct sum above is topological. Further, P_M is a *retraction* of X onto M and is therefore an *identification*. Consider the fibre structure (bundle) (X, M, P_M). We claim that it is *locally trivial*, i.e., that for every $m \in M$, there exists a neighborhood U with $x \in U \subset M$ and a map (slicing map) $\sigma_U: U \times P_M^{-1}(U) \to X$ with the properties that

(a) $P_M \circ \sigma_U(m, v) = m$ if $m \in U$; $v \in P_M^{-1}(U)$ and

(b) $\sigma_U \circ (P_M(v), v) = v$ if $v \in P_M^{-1}(U)$.

Indeed, we may take $U = M$ and $\sigma_U(m, v) = m + v - P_M(v)$. Since M is paracompact this implies (e.g. [36, Ch. 2]) that the fibre structure (X, M, P_M) is a (regular) *fibration*; in other words, the covering homotopy condition holds for every metric space and hence for every topological space.

We next observe that $\ker P_M$ will serve as a *fibre* for the bundle (X, M, P_M). Indeed the fibre over m, $P_M^{-1}(m)$, is simply a translate of $\ker P_M$, namely, $P_M^{-1}(m) = m + \ker P_M$. Therefore, the family $(X, M, \ker P_M, P_M)$ forms a *fibre bundle* with base space M, total space X, fibre $\ker P_M$ and bundle projection P_M. Moreover this fibre bundle is *trivial* in the sense that it is equivalent to the product bundle $(M \times \ker P_M, M, \ker P_M, P)$ where $P: M \times \ker P_M \to M$ is projection on the first factor. For the map $\tau(x) = (P_M(x), x - P_M(x))$ is a homeomorphism and we clearly have $P_M = P \circ \tau$. Since projection from a product space to a factor is always open we have

THEOREM 1. *If P_M is continuous, then it is open.*

We mention in passing that in some problems (e.g. [16] and §2 of this present paper) it is convenient to consider certain "normalizations" of P_M. Namely, letting $S = \{x \in X: \|x\| = 1\}$, we define $\psi_M: X \sim M \to S \cap \ker P_M$ by $\psi_M(x) = \|x - P_M(x)\|^{-1}(x - P_M(x))$. We also define the *equi-distant manifold of* M by $\text{e.d.m}(M) = \{x: \|x - P_M(x)\| = 1\}$. We then have the following conditions for the continuity of P_M.

THEOREM 2. *The following properties of best approximation are equivalent:*

(a) P_M *is continuous on* X*;*

(b) ψ_M *is continuous on* $X \sim M$*;*

(c) $P_M|$ e.d.m. (M) *is continuous on* e.d.m. (M).

The straightforward proof is omitted.

In [28], M. Nashed obtains a decomposition of a Hilbert space H by a closed convex subset $K \subset H$ as follows: every $u \in H$ can be written as the sum of some $x \in H$ and $P_K(x) \in K$, with x depending continuously on u. Since P_K is continuous (even Lipschitzian) this result can be expressed by the statement that $I + P_K$ is a self homeomorphism of H (I is the identity operator on H). Nashed raised the question as to whether this kind of decomposition obtains in certain non-Hilbert spaces. We now observe that the answer is clearly in the affirmative if we restrict K to be a linear variety with a continuous BAO.

THEOREM 3. *Let* X *be a Banach space,* M *a Chebyshev subspace of* X, *and assume* P_M *continuous. Then* $I + P_M$ *is a homeomorphism of* X *with itself.*

Proof: The operator $I - \frac{1}{2}P_M$ is a continuous 2-sided inverse for $I + P_M$.

With these preliminaries out of the way let us turn now to a consideration of $\ker P_M$. For the moment we do not assume P_M continuous. It is clear that the kernel is closed and consists of all $x \in X$ for which $\operatorname{dist}(x, M) = \|x\|$. Such x have been called *normal* to M [12] and *orthogonal* to M [19]. Since P_M is homogeneous, it follows that $\ker P_M$ is symmetric about θ and is a union of rays emanating from θ. (In the terminology of [38], $\ker P_M$ is the *projecting cone* of the set $S \cap$ e.d.m. (M).) As such, $\ker P_M$ is star-shaped with respect to θ and hence *contractible.* Also, $\ker P_M$ is nowhere dense in X.

It is possible to characterize the elements of $\ker P_M$ by means of linear functionals. This was first done by Murray [27, App. I] and later by Singer [27, §1] in an equivalent form. See also James [19, §2]. Let $S^* =$

$\{f \in X^*: \|f\| = 1\}$ and $M^{\perp} = \{f \in X^*: f(M) = 0\}$ (the annihilator of M). Then according to Murray et. al., $x \in \ker P_M$ *iff there exists* $f \in S^* \cap M^{\perp}$ *such that* $f(x) = \|x\|$. The necessity of this condition requires the Hahn-Banach theorem. For further results of this nature see Singer [34] and the references cited there.

It is easy to verify that P_M is continuous on X iff it is continuous at each point of $\ker P_M$ [6, 15]. Hence we may assert that the operator $P_M(\cdot)$ is continuous iff the functional $\|P_M(\cdot)\|$ is continuous.

THEOREM 4. *If* P_M *is continuous, then* $\ker P_M$ *is a strong deformation retract of* X.

Proof: Define a homotopy h_t by $h_t = I - tP_M$ for $0 \leqq t \leqq 1$. Then $h_0 = I$, $h_1 = I - P_M$, a retraction of X onto $\ker P_M$, and $x \in \ker P_M$ implies $h_t(x) = x$ for all t, qed.

It is not known whether this result continues to hold when P_M is discontinuous. However, we do have the following result. We use the standard abbreviations AR for absolute retract and ANR for absolute neighborhood retract.

THEOREM 5. $\ker P_M$ *is a strong deformation retract of* X *iff it is a neighborhood retract in* X.

Proof: A known result in retract theory is that a closed subset ($\ker P_M$) of an AR(X) is an AR iff it is a strong deformation retract (of X) [17, Ch. II, 7.10]. Thus if $\ker P_M$ is a strong deformation retract, then it is actually an ANR since X is an ANR. Conversely, if $\ker P_M$ is a neighborhood retract in X, then it is an ANR. Being contractible, it is an AR, qed.

The next result is basic in studying the relation between the continuity of P_M and the topology of $\ker P_M$. We let Q_M be the quotient map: $X \to X/M$ defined as usual by $Q_M(x) = x + M$. We will say that an operator from one Banach space into another is *weakly continuous* if it is continuous when both spaces are given their weak topologies.

THEOREM 6. P_M *is continuous iff* $Q_M | \ker P_M$ *is a homeomorphism of* $\ker P_M$ *with* X/M.

Proof: We first show that $Q_M | \ker P_M$ (abbreviated to Q for the remainder of this proof) is a continuous and weakly continuous *bijection* of $\ker P_M$ onto X/M. The continuity is trivial and Q is weakly continuous because it is a restriction of a bounded linear operator Q_M and any such operator is weakly continuous. Next, Q is surjective because any $x + M \in X/M$ can be written as $(x - P_M(x)) + M$ and $x - P_M(x) \in \ker P_M$. Also, Q is injective because of the direct sum decomposition of X by M and $\ker P_M$ mentioned earlier in this section. Thus if $Q(x) = Q(y)$, where x and $y \in \ker P_M$ then $x - y \in M$ and so $x = y$. Actually, the converse is also true, so that Q is injective iff M is Chebyshev. Up to this point we have not needed to assume any continuity on the part of P_M. Suppose that Q^{-1} is continuous on X/M. Then P_M is continuous. For let $\|x_n - x\| \to 0$. Then $\operatorname{dist}(x_n - x, M) \to 0$ and so $\|P_M(x_n) - P_M(x)\| \leqq \|P_M(x_n) - P_M(x) - (x_n - x)\| + \|x_n - x\| = \|Q^{-1}(x) - Q^{-1}(x_n)\| + \|x_n - x\| \to 0$.

Finally, assume that P_M is continuous. Let $x_n + M \to x + M$ in X/M and let $\varepsilon > 0$. Since P_M is continuous at $Q^{-1}(x + M)$ there is $\delta > 0$ so that $\|z - Q^{-1}(x + M)\| < \delta$ implies $\|Pz\| < \frac{1}{2}\varepsilon$. Let $V = \{z \in X: \|z - Q^{-1}(x + M)\| < \min(\delta, \frac{1}{2}\varepsilon)\}$. Now $Q_M(V)$ is open and contains $x + M$ so contains $x_n + M$ if n is sufficiently large. For each such n let $z_n \in V$ be such that $Q_M(z_n) = x_n + M$. But $Q_M(z_n) = z_n + M$; thus there is $m_n \in M$ such that $x_n - z_n = m_n$. Hence $P_M(x_n) = m_n + P_M(z_n) = x_n - z_n + P_M(z_n)$. Therefore,

$$
\begin{aligned}
\|Q^{-1}(x_n + M) - Q^{-1}(x + M)\| &= \|x_n - P_M(x_n) - (x - P_M(x))\| \\
&\leqq \|x_n - P_M(x_n) - z_n\| + \|z_n - (x - P_M(x))\| = \\
&= \|P_M(z_n)\| + \|z_n - (x - P_M(x))\| < \tfrac{1}{2}\varepsilon + \tfrac{1}{2}\varepsilon = \varepsilon , \qquad \text{qed}
\end{aligned}
$$

Let us note that Theorem 6 can be rephrased as follows: P_M is continuous iff $x_n, x \in \ker P_M$ and $\operatorname{dist}(x_n - x, M) \to 0$ implies $\|x_n - x\| \to 0$.

We will discuss the analogous result for the *weak* continuity of P_M later in this section. In keeping with our theme of studying smoothness properties of P_M by means of structural properties of $\ker P_M$ we mention that the following results can be proved in an essentially similar manner.

THEOREM 6a. *P_M is linear iff $Q_M | \ker P_M$ is an isometry between $\ker P_M$ and X/M.*

THEOREM 6b. *P_M satisfies a Lipschitz condition on X iff $Q_M | \ker P_M$ is a Lipschitz equivalence between $\ker P_M$ and X/M.*

(We recall that a Lipschitz equivalence, Q, between two metric spaces is a bijection such that both Q and Q^{-1} are Lipschitz continuous on their respective domains.)

It is, of course, possible to draw various inferences from Theorem 6, and we mention a few by way of illustration; other applications of this theorem occur in §2 below.[3] First we note that if P_M is continuous, then $\ker P_M$ is *locally* contractible. If $\operatorname{codim} M < \infty$, then $\{x : x = \psi_M(x)\}$ is a $(\operatorname{codim} M - 1)$-sphere. Also $\ker P_M$ is C^∞-Banach *manifold* modeled on X/M (for infinite-dimensional differentiable manifolds see [9, §4] and the references cited there). We would like to think of $\ker P_M$ as a submanifold of X. To this end we have the following result.

THEOREM 7. *Assume that P_M is continuous and that M is complemented in X, that is, there exists a bounded linear projection P of X onto M. Let $N = P^{-1}(\theta)$ so that $M \oplus N = X$. Then $\ker P_M$ is a properly imbedded C^0-submanifold of X, modeled on N.*

Proof: X is trivially a C^∞-manifold modeled on itself. We want fo find a covering of $\ker P_M$ by domains of charts (U_α, ϕ_α) (for X) such that $\phi_\alpha(U_\alpha \cap \ker P_M) = \phi_\alpha(U_\alpha) \cap N$. Now N and X/M are (linearly) homeomorphic so there exists a homeomorphism ψ of $\ker P_M$ with N. As a chart

---

[3] In a private communication P. Morris has shown how our Theorem 6 can be applied to considerably shorten the proof of the theorem stated in footnote 2.

for X consider (X, ϕ) where $\phi(x) = P_M(x) + \psi(x - P_M(x))$. This chart is clearly C^0-related to the chart (X, I) and we have $\phi(\ker P_M) = N$, qed.

REMARK. M will of course be complemented in X if min(dim M, codim M) $< \infty$. In other cases, results of Lindenstrauss [26, §3 and 24, Chapter II] imply that if P_M is *uniformly* continuous on X (which happens iff P_M is Lipschitz continuous on X [16]) and if M is complemented in M** (which happens iff M is complemented in *some* dual space) then M is complemented in X.

Unfortunately, it seems unlikely that $\ker P_M$ can be properly imbedded as a C^k-submanifold of X for $k \geqq 1$. The reason is that P_M is generally *not* continuously Fréchet differentiable on the open set $X \sim M$. Even a fairly "nice" space such as ℓ^p contains M for which P_M fails to be even Gateaux differentiable. A discussion of this problem is contained in [16].

THEOREM 8. *If* dim M $< \infty$ *then* $\ker P_M$ *is homeomorphic with* X.

Proof: This is another consequence of Theorem 6 together with some known results from infinite-dimensional homeomorphism theory. The key fact is that a Banach space is homeomorphic with its unit sphere (= S). This follows from Lemma 4.1 of [23] and the fact that all separable infinite-dimensional Banach spaces are homeomorphic. Since every hyperplane of a Banach space is also known to be homeomorphic to the unit sphere (e.g., [23, Th. 2.4]), we conclude that X is homeomorphic to all its closed subspaces of finite codimension. Thus X is homeomorphic with any complementary subspace of M and so, by Theorem 6, X is homeomorphic with $\ker P_M$, qed.

While on the subject of topology and finite-dimensional Chebyshev subspaces it is appropriate to mention the following theorem due to Gohberg and Krein [13, §1]: *let* M *and* N *be subspaces with* dim M < dim N *and* M *Chebyshev. Then* $S \cap N \cap \ker P_M \neq \phi$. The proof makes essential use of the Borsuk antipodal theorem.

As a final direct consequence of Theorem 6 we give without proof

THEOREM 9. *If* $\operatorname{codim} M < \infty$ *then* P_M *is continuous iff* $s \cap \ker P_M$ *is compact.*

This theorem is essentially known since Cheney and Wulbert [6, Prop. 10] have proven the obviously equivalent result: if $\operatorname{codim} M < \infty$ then P_M is continuous iff $\ker P_M$ is *boundedly compact.*

In the remainder of this section we consider the problem of the weak sequential continuity of BAO's. We note that conditions on X have been given by Klee [22, §2] which guarantee the weak continuity of P_M if, in particular $\dim M < \infty$.

THEOREM 10. (a) *If* $Q_M | \ker P_M$ *is a weak homeomorphism of* $\ker P_M$ *with* X/M, *then* P_M *is weakly continuous.*

(b) *If* $\operatorname{codim} M < \infty$ *and* P_M *is continuous, then it is weakly continuous.*

Proof of (a). Again write $Q = Q_M | \ker P_M$. Let $x_n \rightharpoonup x$ ("$\rightharpoonup$" denotes weak convergence). Write $x_n = m_n + k_n$ and $x = m + k$ where $m_n, m \in M$ and $k_n, k \in \ker P_M$. Since $(X/M)^* = M^{\perp}$ we have $k_n + M \rightharpoonup k + M$; hence $k_n \rightharpoonup k$ because Q has a weakly continuous inverse. Therefore, $P_M(x_n) = m_n \rightharpoonup m = P_M(x)$.

Proof of (b). By (a) it suffices to show that Q is a weak homeomorphism. Suppose $k_n, k \in \ker P_M$ and $k_n + M \rightharpoonup k + M$. Since $\dim(X/M) < \infty$, $k_n + M \to k + M$ whence by Theorem 6, $\|k_n - k\| \to 0$, and *a fortiori*, $k_n \rightharpoonup k$, qed.

THEOREM 11. *Assume* X *is reflexive. Then the following are equivalent:*

(a) P_M *is weakly continuous;*

(b) $\ker P_M$ *is weakly closed;*

(c) $Q_M | \ker P_M$ *is a weak homeomorphism.*

Proof: It suffices to prove that (b) $\Longrightarrow$ (c). So assume (b) and let k_n, $k \in \ker P_M$ with $k_n + M \rightharpoonup k + M$. Let k_{n_i} be any weakly convergent subsequence of $\{k_n\}$, say $k_{n_i} \rightharpoonup \ell$; we have $\ell \in \ker P_M$. Now, in X/M we have

$k_{n_i} + M \rightharpoonup$ both $k + M$ and $\ell + M$. Hence $k + M = \ell + M$, so, as usual, $k = \ell$. We have shown that $k_n \rightharpoonup k$, qed.

This result is the analogue of Theorem 6 for weakly continuous best approximation in reflexive spaces. We note the following *corollary: if* X *is reflexive and* codim M $< \infty$ *then* P_M *is weakly continuous.* To prove this, it is enough to show that $\ker P_M$ is weakly closed. Suppose $k_n \in \ker P_M$ and $k_n \rightharpoonup z$. There exist $f_n \in S^* \cap M^{\perp}$ such that $f_n(k_n) = \|k_n\|$ and since $\dim M^{\perp} < \infty$, we may assume that $\|f_n - f\| \to 0$ for some $f \in S^* \cap M^{\perp}$. Then

$$\|k_n\| = f_n(k_n) \to f(z) \leqq \|z\| \leqq \underline{\lim}\, \|k_n\|$$

and so $f(z) = \|z\|$, whence $z \in \ker P_M$. Note that no use was made of reflexivity to prove weak closure of $\ker P_M$ for codim M $< \infty$.

Our final result of this section demonstrates the existence of (non-Hilbert) Banach spaces with the property that *all* P_M are weakly continuous. It is very probable that there are many more Banach spaces enjoying this property beside the ones we give here.

Let X be smooth. For $x \neq \theta$ we define

$$G(x, y) = \lim_{t \to 0+} \frac{\|x + ty\| - \|x\|}{t} \cdot \text{ where } G(x, y)$$

is the directional (or Gateaux) derivative of the norm at x in the direction y. It is well known that $G(x, y)$ is a bounded linear functional in y and, in fact, that $G(x, \cdot) \in S^*$. Let us suppose that the map $x \to G(x, \cdot)$ is weakly continuous. Then if M is a one-dimensional Chebyshev subspace of X, $\ker P_M$ is weakly closed. For suppose $P_M(k_n) = \theta$ and $k_n \rightharpoonup z \neq \theta$. Now $G(k_n, \cdot) \rightharpoonup G(z, \cdot)$ and $G(k_n, m) = 0$, where span $\{m\} = M$. Thus $G(z, m) = 0$ and $G(z, \cdot) \in S^* \cap M^{\perp}$. Since $G(z, z) = \|z\|$, we must have $z \in \ker P_M$.

THEOREM 12. *Let* X *be reflexive, smooth, and strictly convex, and assume that the directional derivative map from* $X \sim \{\theta\}$ *to* S^* *is weakly continuous. Then* P_M *is weakly continuous for all* $M \subset X$.

Proof: By hypothesis every M is Chebyshev. Choose any M. By Theorem 11, it will suffice to prove that $\ker P_M$ is weakly closed. Let $\{m_t\}$ represent an indexing of the vectors in $S \cap M$, so that $M = \overline{\text{span}}\{m_t\}$. By the kernel intersection theorem of [16] $\ker P_M = \bigcap_t \ker P_{M_t}$, where $M_t = \text{span}\{m_t\}$. But by the preceding remarks, each $\ker P_{M_t}$ is weakly closed, and therefore $\ker P_M$ is too, qed.

It has been implicitly observed by Browder [4, §4] that the ℓ^p spaces, $1 < p < \infty$, satisfy the assumptions of Theorem 12. On the other hand, he proved that some L^p spaces do not possess *any* weakly continuous duality mappings.[4] So even for these spaces the problem of weak continuity of BAO's is open at this time.

§2. *Duality Mappings and Approximation Theory.*

As background references for this section we give [3, 5, 7, 8, 30, 37]. The symbols θ, X, M, X*, S, S*, M, X/M, P_M and Q_M will continue to have the same meaning as in §1. Following [7] we define the following properties which a Banach space X may or may not possess; if P is any such property, we write "X is (P)" to mean X has property (P).

Definition. X is (R) (*rotund* = strictly convex) iff S contains no line segments. X is (G) (smooth) iff the norm is Gateaux differentiable at every $x \neq \theta$. X is (F) iff the norm is *Fréchet differentiable* at every $x \neq \theta$. X is (H) iff X is (R) and $x_n \to x$, $\|x_n\| \to \|x\| \implies \|x_n - x\| \to 0$.

It is well known [7] that for reflexive spaces X, properties (R) and (G), and properties (H) and (F) are dual, that is, X is (R) (resp. (H)) iff X* is (G) (resp. (F)).

Let us consider property (F) in more detail. By definition, if X is (F), then for $x \neq \theta$,

---

[4] Opial (*Bull. AMS, 73* (1967), 591-597) has shown that *no* $L^p[0, 2\pi]$ *space, where* $1 < p < \infty$, $p \neq 2$, possesses any weakly continuous duality mappings.

$$\lim_{t \to 0} \frac{\|x + ty\| - \|x\|}{t} = G(x, y)$$

is approached uniformly as y varies in S. This is equivalent to saying that $\|x + y\| = \|x\| + G(x, y) + 0(\|y\|)$. As we have already mentioned, $G(x, \cdot) \in S^*$, since $G(x, x) = \|x\|$ and $|G(x, y)| \leqq \|y\|$. So, $G(x, \cdot)$ defines a normalized support functional [30] to the unit cell of X at the point $x/\|x\|$. Let $\mu(x) = G(x, \cdot)$ so that $\mu: X \sim \{\theta\} \to S^*$; recall that μ is *demicontinuous* iff it is continuous from the norm topology on X to the weak star topology on X*. Then it is known [30] that μ is demicontinuous iff X is (G) and that μ is continuous iff X is (F). The map μ will be called the Fréchet derivative of $\| \cdot \|$.

Suppose that X is (F); we define a variant of the map μ which will will prove useful in what follows. Let $\nu: X \to X^*$ be defined by

$$\begin{aligned} \nu(x) &= \|x\| \mu(x) \text{ if } x \neq \theta, \\ &= \theta, \text{ if } x = \theta . \end{aligned}$$

Clearly ν is continuous and has the property that

$$<\nu(x), x> = \|\nu(x)\| \, \|x\| = \|x\|^2 ,$$

where we have used the notation $<f, x>$ for f(x) if $f \in X^*$, $x \in X$. The map ν has been extensively studied by Cudia [8] where it is called the (extended) *spherical image map*. ν also satisfies the requirements of a *duality map,* in the terminology of Browder [3].

For the remainder of this section we will assume that X is reflexive and rotund, unless otherwise stated. Then X* is reflexive and smooth. We will consistently denote by T the (demicontinuous) spherical image map from X* to X** = X; explicitly, if $f \in X^*$ is not θ, then T(f) = x where $f(x) = \|f\| \, \|x\| = \|x\|^2$, while $T(\theta) = \theta$. Also, if M is a closed (Chebyshev) subspace of X, then T_M will be the restriction of T to $M^{\perp}$: $T_M = T \mid M^{\perp}$.

THEOREM 13. *If* T *is continuous at* $f_0 \in S^* \cap M^\perp$ *then* P_M *is continuous at all points of the set* $\psi_M^{-1}(T(f_0))$.

Proof: This is essentially a permutation of some results of Smulian [37] and Cudia [8] which together imply that the continuity of T at $f_0 \in S^*$ is equivalent to the condition that whenever $x_n \in S$ and $f_0(x_n) \to 1$ then $\|x_n - T(f_0)\| \to 0$. Thus if $\|y_n - y_0\| \to 0$ where $\psi_M(y_0) = T(f_0)$, then because X is reflexive, it follows that $\psi_M(y_n) \rightharpoonup T(f_0)$ and so $\|\psi_M(y_n) - T(f_0)\| \to 0$. Then also $\|P_M(y_n) - P_M(y_0)\| \to 0$ (by the proof of Theorem 2), qed.

Notice that in this proof we tacitly employed the fact that range $(T_M) = \ker P_M$; it is further true that $T^{-1}(\ker P_M) = M^\perp$ if X is smooth. We also remark that, in view of the reflexivity assumption, T_M will be $1-1$ if X/M is smooth.

The dual space $(X/M)^*$ is well-known to be (congruent to) $M^\perp$; because of the assumed reflexivity we also have $(M^\perp)^* = X/M$. Let $T_{/M}$ denote the spherical image map from the smooth space $M^\perp$ to its dual X/M. It is easy to see that $T_{/M}(f) = T(f) + M$ for $f \in M$. This means that the following diagram is *commutative:*

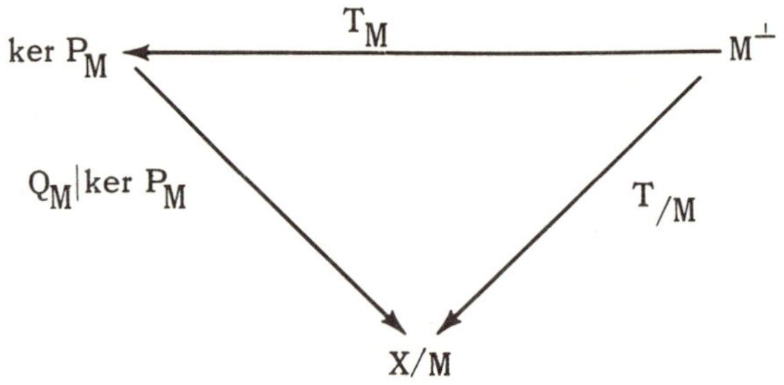

We now come to the main result of this section.

THEOREM 14. *Let* X *be rotund and reflexive and assume* codim $M < \infty$. *Then* P_M *is continuous if and only if* T_M *is continuous (when both* $M^\perp$ *and* $\ker P_M$ *are given the norm topology).*

Proof: The proof will be based on some results about monotone operators. Namely, being duality maps, both T and T/M are *maximal monotone operators* as defined by Browder (e.g., [3]). Now if U is a maximal monotone map from a reflexive space Y to Y*, then it has been shown by Browder [3, Lemma 1.2] that $y_n \to y$ and $\|U(y_n) - g\| \to 0$ together imply $g = U(y)$. We may paraphrase this by stating that U is a *closed mapping* from Y with the weak topology to Y* with the norm topology. Now $M^{\perp}$ and X/M are by assumption finite dimensional. We may thus invoke a theorem due to Kato [20] and Browder [2] to the effect that if U is a locally bounded [20] demicontinuous monotone map from Y to Y* and dim Y $< \infty$ then U is continuous. Thus we may assert that $T_{/M}$ is a closed and continuous map from $M^{\perp}$ to X/M and in particular is an identification. As before, let $Q = Q_M | \ker P_M$. Now we see that $T_M = Q^{-1} \circ T_{/M}$ and $Q^{-1} = T_M \circ T_{/M}$, whether or not $T_{/M}$ is one-to-one. Suppose that P_M is continuous; then by Theorem 6, Q^{-1} is continuous, and so is T_M. And conversely, if T_M is continuous, then so is Q^{-1} because $T_{/M}$ is an identification; a final appeal to Theorem 6 completes the proof, qed.

COROLLARY. P_M *is continuous for all finite codimensional* $M \subset X$ *iff* T *is continuous on* X* *given the finite topology.*

Proof: We recall that the finite topology for a real linear space Y is the topology which Y gets as the *direct limit* of its finite-dimensional subspaces (each of which has the natural topology). Equivalently, it is the *weak* topology on Y induced by the family of finite-dimensional subspaces. The point is that a function on Y is continuous with respect to the finite topology on Y iff its restriction to each finite-dimensional subspace of Y. is continuous. With these observations, the Corollary is seen to follow directly from the theorem

REMARKS. (a) It is possible to give an alternative and somewhat more direct proof of the theorem using the Cheney-Wulbert criterion (cf. Theorem 9) for the continuity of P_M, in lieu of the Browder-Kato theorem guaranteeing continuity of $T_{/M}$. We have given preference to the proof presented here, however, because it seems to offer the greater hope of generalization (to the case where M is not assumed to be finite codimensional). (b) Clearly, Theorem 14 is equivalent to Theorem 9 and/or the Cheney-Wulbert criterion for reflexive rotund spaces X.

One of the main problems in the area of best approximation out of Chebyshev subspaces is that of determining *necessary* conditions for the continuity of *all* P_M, where M runs through the subspaces of a rotund and reflexive X. We recall from the introduction that for this it is *sufficient* that X* be (F), this condition in turn being possible only if X is reflexive and (R). In [16] the authors conjectured that all P_M are continuous without any additional assumption about X*. Now, however, Theorem 14 together with the fact that X* is (F) iff $T: X^* \to X$ is continuous tends to suggest that perhaps the strong smoothness (F) of X* is in fact necessary for the continuity of all P_M.

In connection with the preceding remarks we mention that (as far as the author knows) there is no known example of a reflexive smooth space that does not have property (F). However, in view of constructions given by Lindenstrauss [25, p. 145] and Singer [35, p. 175] it seems likely that such examples exist.

THEOREM 15. *Assume that* X *is reflexive and* (R) *and that* X/M *is* (F). *Then* P_M *is continuous if* T_M *is continuous.*

The proof follows directly from the method of proof of Theorem 14; the effect of the assumption about X/M being to guarantee that $T_{/M}$ has a continuous inverse, and hence is a homeomorphism if T_M is continuous. We note that X/M will be (F) if in particular X is (F).

The point of this theorem is that we have removed the hypothesis of Theorem 14 that $\operatorname{codim} M < \infty$. However, the assumption that X/M is (F) is still quite strong. We conjecture that in fact P_M is continuous whenever T_M is. An examination of the proof of Theorem 14 shows that this will be true whenever $T_{/M}$ is an identification. We are thus led to the interesting question: when is the spherical image map of a reflexive and strongly smooth space an identification (in the norm topologies)?

The next theorem contains some answers and complements to the above questions.

THEOREM 16: *Let* X *be rotund and reflexive.*

(a) $\ker P_M$ *is* (H) *implies* T_M *is continuous;*

(b) $\ker P_M$ *is* (H) *iff* P_M *is continuous and* X/M *is* (H);

(c) *if all* P_M *are continuous, then* X *is* (H) *iff every* X/M is (H).

Proof of (a): Let $f \in M^{\perp}$ with $\|f_n - f_0\| \to 0$. If $f_0 = \theta$ then $\|T_M(f_n)\| \to 0$. Otherwise let $g_n = \|f_n\|^{-1} f_n$, $n \geqq 0$. Since T_M is positively homogeneous, it will suffice to show $\|T_M(g_n) - T_M(g_0)\| \to 0$. Let $T_M(g_n) = x_n = \psi_M(x_n)$. By the (H) assumption on $\ker P_M$, it is enough to prove $x_n \to x_0$. Let y_0 be any weak sequential limit point of $\{x_n\}$; $\|y_0\| \leq 1$. Since $|g_0(y_0) - g_0(x_0)| \leqq |g_0(y_0) - g_0(x_{n_i})| + |g_0(x_{n_i}) - g_{n_i}(x_{n_i})| + |g_{n_i}(x_{n_i}) - g_0(x_0)| \to 0$, $y_0 = x_0$ because X is (R).

Proof of (b): This follows from (a) and Theorem 6.

Proof of (c): If X is (H) then X* is (F), T is continuous, T_M is continuous, and so $M = (X/M)^*$ is (F). (As has already been noted, X is (H) implies all P_M continuous.) Conversely, it will suffice to show that all T_M are continuous for then T will be continuous and X* will be (F). But $T_M = Q^{-1} \circ T_{/M}$ and both of these are continuous by hypothesis ($Q = Q_M | \ker P_M$).

We have seen (Corollary to Theorem 11) that if X is reflexive and codim $M < \infty$, then P_M is weakly continuous; while if X is not reflexive, then P_M (and also ψ_M by Theorem 2) may be discontinuous even when X is (R) (we are still assuming codim $M < \infty$). Our final result complements these facts.

THEOREM 17. *Let* X *be reflexive,* M *a finite-codimensional Cheby-*

shev subspace with X/M *rotund. Let* x_n, $x \in X \sim M$ *with* $x_n \to x$. *Then* $\psi_M(x_n) \to \psi_M(x)$ *and* $\text{dist}(\psi_M(x_n) - \psi_M(x), M) \to 0$.

Proof: Since P_M is weakly continuous and X is reflexive, it is enough to prove that $\text{dist}(x, M)$ is an accumulation point of every subsequence of $\{\text{dist}(x_n, M)\}$. However, this follows routinely from the formula $\text{dist}(z, M) = \max\{|f(z)| : f \in S^* \cap M^{\perp}\}$ and the compactness of $S^* \cap M^{\perp}$. (It is also to be noted that $\underline{\lim}\, \text{dist}(x_n, M) \geqq \text{dist}(x, M) > 0$.) The second statement of the theorem follows from the first and the fact that X/M is uniformly convex (since it is finite-dimensional) and so has property (H).

We note that in view of a theorem of Klee [21, §3] X/M will be rotund if in particular X is.

In closing, we remark that whenever T_M is *linear* then P_M is linear and X/M is a *Hilbert* space (under the inner product $(x + M, y + M) = \langle {T_{/M}}^{-1}(y + M), x + M \rangle$). Unfortunately, counterexamples exist which prove the converse false.

REFERENCES

[1] J. Blatter, P. Morris, and D. Wulbert, *Continuity of the set-valued metric projection,* to appear.

[2] F. Browder, Continuity properties of monotone nonlinear operators in Banach spaces, *Bull. AMS, 70* (1964), 551-553.

[3] ______, Multivalued monotone nonlinear mappings and duality mappings in Banach spaces, *Trans. AMS, 118* (1965), 338-351.

[4] ______, Fixed point theorems for nonlinear semicontractive mappings in Banach spaces, *Arch. Rat. Mech. and Anals., 21* (1966), 259-269.

[5] C. Buck, Applications of duality in approximation theory, *Proc. of Symposium on Approximation of Functions,* Ed. H. Garabedian, Elsevier, New York, 1965, 27-43.

[6] W. Cheney and D. Wulbert, *The existence and unicity of best approximations,* to appear.

[7] D. Cudia, Rotundity, *Proc. of the Symposium on Convexity,* Ed. V. Klee, Amer. Math. Soc., Providence, 1963, 73-97.

[8] ______, The geometry of Banach spaces. Smoothness, *Trans. AMS, 110* (1964), 284-314.

[9] J. Eells, A setting for global analysis, *Bull. AMS, 72* (1966), 751-807.

[10] N. Efimov and S. Stechkin, series of papers on Chebyshev sets in *Dokl. Akad. Nauk SSSR* from 1958 on.

[11] K. Fan and I. Glicksberg, Some geometric properties of the spheres in a normed linear space, *Duke Math. J., 25* (1958), 553-568.

[12] R. Fortet, Remarques sur les espaces uniformément convexes, *Bull. Soc. Math. France, 69* (1941), 23-45.

[13] I. Gohberg and M. Krein, The basic propositions on defect numbers, root numbers and indices of linear operators, *AMS Translations* (Series 2), *Vol. 13,* 185-265.

[14] R. Hirshfeld, On best approximation in normed vector spaces, II, *Nieuw Archief voor Wisk. (3), VI* (1958), 99-107.

[15] R. Holmes, Dissertation, MIT, Cambridge, 1964.

[16] R. Holmes and B. Kripke, Smoothness of approximation, to appear in *Michigan Math. J.*

[17] S. Hu, *Theory of Retracts,* Wayne State Univ. Press, Detroit, 1965.

[18] R. James, Orthogonality in normed linear spaces, *Bull. AMS, 53* (1947), 559-566.

[19] ______, Orthogonality and linear functionals in normed linear spaces, *Trans. AMS, 61* (1947), 265-292.

[20] T. Kato, Demicontinuity, hemicontinuity, and monotonicity, *Bull. AMS, 70* (1964), 548-550.

[21] V. Klee, Some new results on smoothness and rotundity in normed linear spaces, *Math. Ann., 139* (1959), 51-63.

[22] ______, Convexity of Chebyshev sets, *Math. Ann., 142* (1961), 292-304.

[23] B. Lin, Two topological problems concerning infinite dimensional normed linear spaces, *Trans. AMS, 114* (1965), 156-175.

[24] J. Lindenstrauss, Extension of compact operators, *Mem. AMS*, no. 47, 1964.

[25] ______, On operators which attain their norm, *Israel J. Math., 1* (1963), 139-148.

[26] ______, On nonlinear projections in Banach spaces, *Mich. Math. J., 11* (1964), 263-287.

[27] F. Murray, Analysis of linear transformations, *Bull. AMS, 48* (1942), 76-93.

[28] M. Nashed, *A decomposition relative to convex sets*, to appear.

[29] R. Phelps, Unique Hahn-Banach extensions and unique best approximation, *Trans. AMS, 95* (1960), 238-255.

[30] G. Restrepo, Differentiable norms in Banach spaces, *Bull. AMS, 70* (1964), 413-414.

[31] W. Rudin and K. Smith, Linearity of best approximation: a characterization of ellipsoids, *Nederl. Akad. Wetensch. Proc. Ser. A 64* (1961), 97-103.

[32] I. Singer, Characterization des elements de meilleure approximation dans un espace de Banach quelconque, *Acta Sci. Math. Szeged, 17* (1956), 181-189.

[33] ______, On the set of best approximations of an element in a normed linear space, *Rev. Math. Pures et App., 5* (1960), 383-402.

[34] ______, On the extension of continuous linear functionals and best approximation in normed linear spaces, *Math. Ann., 159* (1965), 344-355.

[35] ______, Some remarks on approximative compactness, *Rev. Math. Pures et App., 9* (1964), 167-177.

[36] E. Spanier, *Algebraic Topology,* McGraw-Hill, New York, 1966.

[37] V. Smulian, Sur la derivabilité de la norme dans l'espace de Banach, *Dokl. Akad. Nauk SSSR, 27* (1940), 643-648.

[38] R. Wijsman, Convergence of sequences of convex sets, cones, and functions, *Bull. AMS, 70* (1964), 186-188.

[39] D. Wulbert, *Structure of Chebyshev sets,* to appear.

Air Force Institute of Technology

SOME SELF-DUAL PROPERTIES OF NORMED LINEAR SPACES*

ROBERT C. JAMES

The principal purpose of this paper is to investigate some self-dual properties of normed linear spaces. The negations of these properties will be grouped in sets designated by P_1, P_2, P_3, P_4 and S. For complete spaces, each of these properties is stronger than reflexivity. All of them are possessed by any normed linear space that is isomorphic to a space with a uniformly convex unit ball, but it is not known whether some one of these properties is equivalent to the existence of such an isomorphism. The results can be indicated by the following diagram:

$$P_1^\infty \Longrightarrow [P_2^\infty, P_3^\infty, P_4^\infty, \sim R] \Longrightarrow [P_1, P_2, P_3, P_4] \Longrightarrow S \Longrightarrow \sim U .$$

The first two implications are not equivalences, but for neither of the last two is it known whether they are equivalences. For each set of brackets, all of the properties indicated by any of the included symbols are equivalent. Each property of type P_1, P_2, P_3, P_4 or S is self-dual. The symbol $\sim R$ indicates *the completion of the space is not reflexive* and the symbol $\sim U$ indicates *not isomorphic to any space with a uniformly convex unit ball.* The remaining symbols denote sets of properties as follows:

P_1 (*Finite tree properties*): The four properties consisting of the following property and all variations obtained by replacing the quantifier "for some positive number ε" by "for any ε with $0 < \varepsilon < 2$" or by creating dual properties by replacing each $x_{\varepsilon_1 \cdots \varepsilon_k}$ by a member $f_{\varepsilon_1 \cdots \varepsilon_k}$

* This work was supported in part by National Science Foundation grant number NSF-GP-6653.

of the unit ball of the first conjugate space:

For some positive number ε *and any positive integer* m, *there is a set* $\{x_{\varepsilon_1 \cdots \varepsilon_k} : 1 \leqq k \leqq m\}$ *containing* 2^m *members of the unit ball for which each* ε_i *is* 1 *or* 2 *and for any set* $\{\varepsilon_1, \varepsilon_2, \ldots, \varepsilon_k\}$, *possibly empty,*

$$x_{\varepsilon_1 \cdots \varepsilon_k} = \tfrac{1}{2}(x_{\varepsilon_1 \cdots \varepsilon_k 1} + x_{\varepsilon_1 \cdots \varepsilon_k 2}), \tag{1}$$

$$\|x_{\varepsilon_1 \cdots \varepsilon_k 1} - x_{\varepsilon_1 \cdots \varepsilon_k 2}\| \geqq \varepsilon . \tag{2}$$

P_1^∞ (*Infinite tree properties*): The following property and the dual property obtained from it by replacing each $x_{\varepsilon_1 \cdots \varepsilon_k}$ by a member $f_{\varepsilon_1 \cdots \varepsilon_k}$ of the unit ball of the first conjugate space:

For some positive number ε, *there is a subset* $\{x_{\varepsilon_1 \cdots \varepsilon_k} : 1 \leqq k\}$ *of the unit ball for which each* ε_i *is* 1 *or* 2 *and for any set* $\{\varepsilon_1, \varepsilon_2, \ldots, \varepsilon_k\}$, *possibly empty,*

$$x_{\varepsilon_1 \cdots \varepsilon_k} = \tfrac{1}{2}(x_{\varepsilon_1 \cdots \varepsilon_k 1} + x_{\varepsilon_1 \cdots \varepsilon_k 2}), \tag{3}$$

$$\|x_{\varepsilon_1 \cdots \varepsilon_k 1} - x_{\varepsilon_1 \cdots \varepsilon_k 2}\| \geqq \varepsilon . \tag{4}$$

P_2 (*Finite flatness properties*): The twenty-four properties consisting of the following property and all variations obtained by replacing the quantifier "for some positive number ε" by "for any ε with $0 < \varepsilon < 2$," or by using the given quantifier or the quantifier "for any ε with $0 < \varepsilon < 1$" and replacing the symbol "conv" by "flat" or by "lin" in either place it occurs (except that it may not be replaced by "lin" in both places), or by creating dual properties by replacing each x_k by a member f_k of the unit ball of the first conjugate space:

For some positive number ε *and any positive integer* n, *there is a subset* $\{x_1, \ldots, x_n\}$ *of the unit ball such that, if* $1 \leqq k < n$, *then*

$$\operatorname{dist}(\operatorname{conv}\{x_1, \ldots, x_k\}, \operatorname{conv}\{x_{k+1}, \ldots, x_n\}) \geqq \varepsilon . \tag{5}$$

P_2^∞ (*Infinite flatness properties*): The twenty-four properties consisting of the following property and all variations obtained by replacing the quantifier "for some positive number ε" by "for any ε with $0 < \varepsilon < 1$," or by replacing the symbol "conv" by "flat" or by "lin" in either place it occurs (except that if the second "conv" is replaced by "lin," then the first is to be replaced only by "flat" and $0 < \varepsilon < 1$ is to be replaced by $0 < \varepsilon < \frac{1}{2}$), or by creating dual properties by replacing each x_k by a member f_k of the unit ball of the first conjugate space:

For some positive number ε, *there is a sequence* $\{x_1, x_2, \ldots\}$ *of members of the unit ball such that, for all* k,

$$\text{(6)} \qquad \operatorname{dist}(\operatorname{conv}\{x_1, \ldots, x_k\}, \operatorname{conv}\{x_{k+1}, \ldots\}) \geqq \varepsilon .$$

P_3 (*Finite basic sequence properties*): The eight properties consisting of the following property and all variations obtained by replacing the quantifier "for some positive number ε" by "for any ε with $0 < \varepsilon < 1$," or by replacing the symbol "conv" by "flat," or by creating dual properties by replacing each x_k by a member f_k of the unit ball of the first conjugate space:

For some positive number ε *and any positive integer* n, *there is a subset* $\{x_1, \ldots, x_n\}$ *of the unit ball for which* $\|u\| \geqq \varepsilon$ *if* $u \in \operatorname{conv}\{x_1, \ldots, x_n\}$ *and, for any positive integer* $k < n$ *and any numbers* $\{a_i\}$,

$$\text{(7)} \qquad \Big\| \sum_{i=1}^{n} a_i x_i \Big\| \geqq \tfrac{1}{2}\varepsilon \Big\| \sum_{i=1}^{k} a_i x_i \Big\| .$$

P_3^∞ (*Basic sequence properties*): The eight properties consisting of the following property and all variations obtained by replacing the quantifier "for some positive number ε" by "for any ε with $0 < \varepsilon < 1$," or by replacing the symbol "conv" by "flat," or by creating dual properties by replacing each x_k by a member f_k of the unit ball of the first conjugate space:

For some positive number ε, *there is a sequence* $\{x_n\}$ *of members of the unit ball for which* $\|u\| \geqq \varepsilon$ *if* $u \in \operatorname{conv}\{x_n\}$ *and for any positive integers* k *and* n *with* $k < n$ *and any numbers* $\{\alpha_i\}$,

$$\left| \sum_{i=1}^{n} \alpha_i x_i \right| \geqq \tfrac{1}{2}\varepsilon \left| \sum_{i=1}^{k} \alpha_i x_i \right| . \tag{8}$$

P_4: The eight properties consisting of the following property and all variations obtained by replacing the quantifier "for some positive number ε" by "for any ε with $0 < \varepsilon < 1$," or by replacing $(f_k, x_i) \geqq \varepsilon$ by $(f_k, x_i) = \varepsilon$, or by creating dual properties by letting x_i and f_i be members of the unit balls of the first and second conjugate spaces, respectively:

For some positive number ε *and any positive positive integer* n, *there is a subset* $\{x_1, \ldots, x_n\}$ *of the unit ball and a subset* $\{f_1, \ldots, f_n\}$ *of the unit ball of the first conjugate space for which*

$$(f_k, x_i) \geqq \varepsilon \text{ if } k \leqq i, \quad (f_k, x_i) = 0 \text{ if } k > i . \tag{9}$$

P_4^∞: The eight properties consisting of the following property and all variations obtained by replacing the quantifier "for some positive number ε" by "for any ε with $0 < \varepsilon < 1$," or by replacing $(f_k, x_i) \geqq \varepsilon$ by $(f_k, x_i) = \varepsilon$, or by creating dual properties by letting x_i and f_i be members of the unit balls of the first and second conjugate spaces, respectively:

For some positive number ε, *there is a sequence* $\{x_1, x_2, \ldots\}$ *of members of the unit ball and a sequence* $\{f_1, f_2, \ldots\}$ *of members of the unit ball of the first conjugate space for which*

$$(f_k, x_i) \geqq \varepsilon \text{ if } k \leqq i, \quad (f_k, x_i) = 0 \text{ if } k > i . \tag{10}$$

S: The two properties: "*Not isomorphic to a space with a uniformly nonsquare unit ball*" and "*the first conjugate space is not isomorphic to a space with a uniformly nonsquare unit ball*" [definitions are given in the proof of Theorem 7].

Now the implications and equivalences described in the introductory paragraph will be established by proving the following Theorems 1 through 7.

THEOREM 1. *The two properties of type* P_1^∞ *are not equivalent and neither is possessed by all nonreflexive Banach spaces.*

Proof: To prove the theorem, it will be shown that the Banach space (c_0) has the first property of type P_1^∞, but that the conjugate ℓ^1 of (c_0) does not have this property. Thus (c_0) does not have the dual property, so the two properties of type P_1^∞ are not equivalent. Since ℓ^1 does not have the first property and (c_0) does not have the dual property, neither property is possessed by all Banach spaces.

To show that (c_0) has the first property of type P_1^∞, we merely let

$$x_{\varepsilon_1 \cdots \varepsilon_k} = (2\varepsilon_1 - 3,\ 2\varepsilon_2 - 3, \ldots,\ 2\varepsilon_k - 3,\ 0,\ 0,\ 0,\ \ldots),$$

and note that $\|x_{\varepsilon_1 \cdots \varepsilon_k}\| = 1$, $\frac{1}{2}(x_{\varepsilon_1 \cdots \varepsilon_k 1} + x_{\varepsilon_1 \cdots \varepsilon_k 2})$ equals

$$\tfrac{1}{2}[(2\varepsilon_1 - 3, \ldots, 2\varepsilon_k - 3, -1, 0, 0, \ldots) + (2\varepsilon_1 - 3, \ldots, 2\varepsilon_k - 3, +1, 0, 0, \ldots)]$$

$$= (2\varepsilon_1 - 3, \ldots, 2\varepsilon_k - 3, 0, 0, 0, \ldots) = x_{\varepsilon_1 \cdots \varepsilon_k},$$

and $\|x_{\varepsilon_1 \cdots \varepsilon_k 1} - x_{\varepsilon_1 \cdots \varepsilon_k 2}\| = 2.$

Now suppose ℓ^1 has the first property of type P_1^∞ and that ε and the subset $S = \{x_{\varepsilon_1 \cdots \varepsilon_k}\}$ of the unit ball have the properties specified in the description of P_1^∞. Let

$$M = \sup\{\|x_{\varepsilon_1 \cdots \varepsilon_k}\| : k \geqq 1 \text{ and } \varepsilon_i = 1 \text{ or } \varepsilon_i = 2 \text{ if } i \leqq k\},$$

and choose a positive integer λ so that there is a member of S for which the sum of the absolute values of the first λ components is greater than $M - \frac{1}{16}\varepsilon$. Then let

$$\text{(11)} \quad M_\lambda = \sup\{\|x_{\varepsilon_1 \cdots \varepsilon_k}\|_\lambda : k \geqq 1 \text{ and } \varepsilon_i = 1 \text{ or } \varepsilon_i = 2 \text{ if } i \leqq k\},$$

where $\| \ \|_\lambda$ denotes the sum of the absolute values of the first λ components. Note that $M_\lambda > M - \frac{1}{16}\varepsilon$. With $x^i_{\varepsilon_1 \cdots \varepsilon_k}$ denoting the ith component of $x_{\varepsilon_1 \cdots \varepsilon_k}$, define $\{m_1, m_2, \ldots, m_\lambda\}$ as follows:

$$m_1 = \lim_{\eta \to 0+} \sup\{x^1_{\varepsilon_1 \cdots \varepsilon_k} : k \geqq 1 \text{ and } \|x_{\varepsilon_1 \cdots \varepsilon_k}\|_\lambda > M_\lambda - \eta\},$$

and, for $1 < j \leqq \lambda$,

$$m_j = \lim_{\eta \to 0+} \sup\{x^j_{\varepsilon_1 \cdots \varepsilon_k} : k \geqq 1,\ \|x_{\varepsilon_1 \cdots \varepsilon_k}\|_\lambda > M_\lambda - \eta, \text{ and } x^i_{\varepsilon_1 \cdots \varepsilon_k} > m_i - \eta \text{ if } i < j\}.$$

Let η be a positive number less than $\varepsilon/(16\lambda)$. Then there are positive numbers $\delta_0 \leqq \delta_1 \leqq \cdots \leqq \delta_\lambda = \eta$ such that if $1 \leqq j \leqq \lambda$ and $\|x_{\varepsilon_1 \cdots \varepsilon_k}\|_\lambda > M_\lambda - \delta_0$ and $x^i_{\varepsilon_1 \cdots \varepsilon_k} > m_i - 3\delta_i$ when $i < j$, then

(12) $$x^j_{\varepsilon_1 \cdots \varepsilon_k} < m_j + \delta_j .$$

Choose $\{e_1, \ldots, e_r\}$ so that

(13) $$\|x_{e_1 \cdots e_r}\|_\lambda > M_\lambda - \tfrac{1}{2}\delta_0 \text{ and } x^j_{e_1 \cdots e_r} > m_j - \delta_j \text{ if } 1 \leqq j \leqq \lambda .$$

Then

(14) $$\sum_{i=\lambda+1}^{\infty} |x^i_{e_1 \cdots e_r}| < \frac{1}{8}\varepsilon ,$$

since otherwise $\|x_{e_1 \cdots e_r}\| > M_\lambda - \frac{1}{2}\delta_0 + \frac{1}{8}\varepsilon > M - \frac{1}{16}\varepsilon - \frac{1}{2}\delta_0 + \frac{1}{8}\varepsilon > M$. It follows from (3) and (13) that there is a value (1 or 2) of α for which $\|x_{e_1 \cdots e_r \alpha}\|_\lambda > M_\lambda - \frac{1}{2}\delta_0$. Then

(15) $$\sum_{i=\lambda+1}^{\infty} |x^i_{e_1 \cdots e_r \alpha}| < \frac{1}{8}\varepsilon ,$$

since otherwise $\|x_{e_1 \cdots e_r \alpha}\| > M_\lambda - \frac{1}{2}\delta_0 + \frac{1}{8}\varepsilon > M - \frac{1}{16}\varepsilon - \frac{1}{2}\delta_0 + \frac{1}{8}\varepsilon > M$. From (3), we have $x^i_{e_1 \ldots e_r} = \tfrac{1}{2}(x^i_{e_1 \ldots e_r \alpha} + x^i_{e_1 \ldots e_r \beta})$, where β is 1 or 2 and $\beta \neq \alpha$. Thus

$$\sum_{i=\lambda+1}^{\infty} |x^i_{e_1\cdots e_r}| \geqq \frac{1}{2}\left[\sum_{i=\lambda+1}^{\infty} |x^i_{e_1\cdots e_r\beta}| - \sum_{i=\lambda+1}^{\infty} |x^i_{e_1\cdots e_r\alpha}|\right],$$

and it follows from (14) and (15) that

$$(16) \qquad \sum_{i=\lambda+1}^{\infty} |x^i_{e_1\cdots e_r\beta}| < \frac{3}{8}\varepsilon .$$

Because of (15) and (16), we can conclude from (4) that

$$(17) \qquad \sum_{i=1}^{\lambda} |x^i_{e_1\cdots e_r\alpha} - x^i_{e_1\cdots e_r\beta}| > \frac{1}{2}\varepsilon .$$

We shall complete the proof by obtaining a contradiction of (17). From (3), (13), and the fact that neither $\|x_{e_1\ldots e_r\alpha}\|_\lambda$ nor $\|x_{e_1\ldots e_r\beta}\|_\lambda$ is larger than M_λ, we have

$$(18) \qquad \|x_{e_1\ldots e_r\alpha}\|_\lambda > M_\lambda - \delta_0 \text{ and } \|x_{e_1\cdots e_r\beta}\|_\lambda > M_\lambda - \delta_0 .$$

Again using (3), but this time with (12) and (13), we start with (18) and obtain successively, for $j = 1, 2, \ldots, \lambda$,

$$x^j_{e_1\cdots e_r\alpha} > m_j - 3\delta_j \text{ and } x^j_{e_1\cdots e_r\beta} > m_j - 3\delta_j .$$

Since (12) implies $x^j_{e_1\cdots e_r\alpha} < m_j + \delta_j$ and $x^j_{e_1\cdots e_r\beta} < m_j + \delta_j$, we have

$$|x^j_{e_1\cdots e_r\alpha} - x^j_{e_1\cdots e_r\beta}| < 4\delta_j < 4\eta ,$$

and

$$\sum_{i=1}^{\lambda} |x^i_{e_1\cdots e_r\alpha} - x^i_{e_1\cdots e_r\beta}| < 4\lambda\eta < \frac{1}{4}\varepsilon ,$$

which contradicts (17).

THEOREM 2. *Each property of type* P_1^∞ *implies* ~R.

Proof: Since the closure of a normed linear space is reflexive if and only if its first conjugate space is reflexive, it is sufficient to consider only the first of the two properties of type P_1^∞. We let X be a normed linear space for which there is a positive number ε and a subset $S = \{x_{\varepsilon_1 \cdots \varepsilon_k}\}$ of the unit ball for which each ε_i is 1 or 2 and conditions (3) and (4) are satisfied. Let $\bar{X}$ be the completion of X and K be the closure in $\bar{X}$ of the convex span of S. A proof by contradiction will be used to show that $\bar{X}$ is not reflexive. If $\bar{X}$ is reflexive, then K is a weakly compact convex subset of a separable subspace of $\bar{X}$ and therefore there is a closed convex subset C of K such that $C \neq K$ and the diameter of $K - C$ is less than $\frac{1}{2}\varepsilon$ (see the lemma in [I] or Theorem 4, page 144 of [V]). Since C is closed and convex, it follows that $C = K$ if $S \subset C$. Therefore, there is an $x_{\varepsilon_1 \cdots \varepsilon_k}$ that belongs to $K - C$. From (3) and (4) and the fact that the diameter of $K - C$ is less than $\frac{1}{2}\varepsilon$, it follows that both $x_{\varepsilon_1 \cdots \varepsilon_k 1}$ and $x_{\varepsilon_1 \cdots \varepsilon_k 2}$ belong to C. Since C is convex, we now have $x_{\varepsilon_1 \cdots \varepsilon_k}$ both belonging to C and not belonging to C.

THEOREM 3. *Each property of type* P_2^∞, P_3^∞ *or* P_4^∞ *is equivalent to* $\sim R$.

Proof: Suppose a normed linear space X has property $\sim R$, i.e., the completion of X is not reflexive. Then it can be shown that for any positive number $\varepsilon < 1$ there is a sequence $\{x_n\}$ of members of the unit ball and a sequence $\{f_n\}$ of members of the unit ball of the first conjugate space for which

$$(f_n, x_k) = \varepsilon \text{ if } n \leqq k, \qquad (f_n, x_k) = 0 \text{ if } n > k .$$

(A proof of this for Banach spaces is given on pages 114-115 of [IV], but the proof given there is valid for normed linear spaces X with the closure of X in X** not equal to X** – i.e., for any X whose completion is not reflexive.) This property and its dual imply all other properties of type P_4^∞. To show that they imply all properties of type P_2^∞, we first note that if $u = \Sigma_1^k \alpha_i x_i$ and $v = \Sigma_{k+1}^p \beta_i x_i$, then

(19) $$f_1(u-v) = \varepsilon \cdot (\Sigma\, \alpha_i + \Sigma\, \beta_i)\,,$$

and

(20) $$f_{k+1}(u-v) = \varepsilon \cdot (\Sigma\, \beta_i)\,.$$

If $\Sigma\, \beta_i = 1$, then $\|u-v\| \geqq \varepsilon$ follows from (20). Thus (6) is satisfied and still is satisfied if the first "conv" is replaced by "flat" or "lin" or the second "conv" is replaced by "flat." If the second "conv" is replaced by "lin," then $\|u-v\| > \tfrac{1}{2}\varepsilon$ follows from (20) if $|\Sigma\, \beta_i| \geqq \tfrac{1}{2}$. If $|\Sigma\, \beta_i| < \tfrac{1}{2}$, then it follows from (19) and $\Sigma\, \alpha_i = 1$ that

$$\|u-v\| \geqq \varepsilon \cdot |\Sigma\, \alpha_i + \Sigma\, \beta_i| \geqq \tfrac{1}{2}\varepsilon\,.$$

The dual properties of type P_2^∞ (those with x_k replaced by f_k) can be proved similarly. To show that ~R implies all properties of type P_3^∞, we note that it follows from ~R that, for any ε with $0 < \varepsilon < 1$, there is a sequence $\{x_n\}$ of members of the unit ball for which $\|u\| \geqq \varepsilon$ if $u \in \text{flat}\{x_n\}$ and, for any positive integers k and n with $k < n$ and any numbers $\{\alpha_i\}$,

$$\|\Sigma_1^n\, \alpha_i x_i\| \geqq \tfrac{1}{2}\varepsilon\, \|\Sigma_1^k\, \alpha_i x_i\|\,.$$

(In the proof of this as given on pages 116-117 of [IV], it is shown that $\|u\| \geqq \Delta$ if $u \in \text{conv}\{x_n\}$ —however, this is done by using a linear functional Φ of unit norm with $\Phi(x_n) = \Delta$ for all n, so the proof actually implies that $\|u\| \geqq \Delta$ if $u \in \text{flat}\{x_n\}$). From this and its dual follow all properties of type P_3^∞.

We have shown that ~R implies all properties of type P_2^∞, P_3^∞ or P_4^∞. We shall now show that each of these properties implies ~R. As we have noted earlier in this proof, each property of type P_4^∞ implies a property of type P_2^∞. Also, each property of type P_3^∞ implies a property of type P_2^∞. The weakest of the properties of type P_2^∞ is the stated one that concludes with inequality (6). That this property implies the completion of the space is not reflexive is given on page 114 of [IV].

THEOREM 4. *For no one of the properties of type* P_1 *is it true that all Banach spaces with that property are non-reflexive.*

Proof: Let B be the product using the Hilbert norm of the finite-dimensional spaces (c_0^n) for $n = 1,2,\ldots$. Then B is reflexive. As in the proof of Theorem 1, if each ε_i is 1 or 2 and $k \leqq n$, then the members

$$x_{\varepsilon_1\cdots\varepsilon_k} = (2\varepsilon_1 - 3,\ 2\varepsilon_2 - 3, \ldots,\ 2\varepsilon_k - 3,\ 0,\ 0,\ 0,\ \ldots,\ 0)$$

of the unit ball of (c_0^n) have the property that

$$x_{\varepsilon_1\cdots\varepsilon_k} = \tfrac{1}{2}(x_{\varepsilon_1\cdots\varepsilon_k 1} + x_{\varepsilon_1\cdots\varepsilon_k 2})$$

and $\|x_{\varepsilon_1\cdots\varepsilon_k 1} - x_{\varepsilon_1\cdots\varepsilon_k 2}\| = 2$. Thus B has the two basic properties of type P_1 (as distinguished from the dual properties). The dual properties are satisfied by the reflexive space that is the product using the Hilbert norm of the finite-dimensional spaces ℓ_n^1 for $n = 1, 2, \ldots$.

THEOREM 5. *If* X *is a normed linear space with property* $\sim$R, *then* X *has all properties of type* P_1, P_2, P_3 *or* P_4 .

Proof: We have proved that X has all properties of type P_2^∞, P_3^∞ and P_4^∞ if X has property $\sim$R [Theorem 3]. Since each property of type P_2^∞, P_3^∞ or P_4^∞ implies the corresponding property of type P_2, P_3 or P_4, we can complete the proof by showing that $\sim$R implies all properties of type P_1 or P_2 that use the quantifier "for any ε with $0 < \varepsilon < 2$." We shall first show that for any ε with $0 < \varepsilon < 2$ there is a subset $\{x_1, \ldots, x_n\}$ of the unit ball such that, if $1 \leqq k < n$, then

$$\text{dist}(\text{conv}\{x_1, \ldots, x_k\},\ \text{conv}\{x_{k+1}, \ldots, x_n\}) \geqq \varepsilon \ . \tag{21}$$

The argument is similar to that used on pages 543-544 of [III]. The details will not be given, since they become clear once the discussion in [III] is understood. We shall only suggest the generalization by noting that if

$n = 4$, then in the notation used in [III] the elements x_1, x_2, x_3 and x_4 belong, respectively, to sets $S(p_1, \ldots, p_{2m}; \{f_i\})$, $S(q_1, \ldots, q_{2m}; \{f_i\})$, $S(r_1, \ldots, r_{2m}; \{f_i\})$, and $S(s_1, \ldots, s_{2m}; \{f_i\})$, where

$$\{p_1, q_1, r_1, s_1, p_2, p_3, q_2, q_3, r_2, r_3, s_2, s_3, p_4, p_5, q_4, q_5, \ldots, \\ s_{2m-2}, s_{2m-1}, p_{2m}, q_{2m}, r_{2m}, s_{2m}\}$$

is an increasing sequence of integers (the author is grateful to D. P. Giesy for suggesting this extension of Theorem 1.1 of [III]). Now let $n = 2^m$ and use the set $\{x_k\}$ of (21) to establish the desired property of type P_1. For each ε_i equal to 1 or 2, let $\xi_{\varepsilon_1 \cdots \varepsilon_m}$ denote x_α where $\alpha - 1$ is the non-negative integer whose digits in the binary system are $\varepsilon_1 - 1, \varepsilon_2 - 1, \ldots, \varepsilon_m - 1$. Then for particular numbers $\{e_1, \ldots, e_k\}$ with $k < m$ and each e_i equal to 1 or 2, define $\xi_{e_1 \cdots e_k}$ to be the average of all $\xi_{\varepsilon_1 \cdots \varepsilon_m}$ which have $\varepsilon_i = e_i$ for $i \leqq k$, so that the set $\{\xi_{\varepsilon_1 \cdots \varepsilon_k}\}$ satisfies (1). It follows from (21) that (2) also is satisfied.

Since the first conjugate space of X is not reflexive if X has property ~R, the preceding arguments can be used for X^* to establish the duals of the properties of type P_1 and P_2 which have been established for X.

The proof of the next theorem will be based on Theorem 5 and the following definition and lemmas.

Definition. For a normed linear space Y to be *finitely represented* in a normed linear space X means that for each finite-dimensional subspace Y_n of Y and each positive number ε there is an isomorphism T_n of Y_n into X for which

$$(1-\varepsilon)\|y\| \leqq \|T_n(y)\| \leqq (1+\varepsilon)\|y\| \quad \text{if } y \in Y_n.$$

LEMMA A. *Let* P *and* Q *be properties possessed by some normed linear spaces. Then* $P \Longrightarrow Q$ *if the following conditions are satisfied:*

(i) *If a normed linear space* X *has property* P, *then there is a normed linear space* Y *with property* ~R *for which* Y *is finitely represented in* X.

(ii) *If* Y *is finitely represented in* X *and has property* ~R, *then* X *has property* Q.

Proof: Suppose a normed linear space X has property P. Use (i) to choose Y which is finitely represented in X and has property ~R. Then it follows from (ii) that X has property Q.

LEMMA B. *If a normed linear space* X *has one of the basic properties of type* P_1, P_2, P_3 or P_4 *(as distinguished from the dual properties), then there is a space* Y *with property* ~R *for which* Y *is finitely represented in* X.

Proof: Each basic property of type P_1, P_2, P_3 or P_4 requires the existence for each n of one or two sets, one of which is a subset $\{x_1^n, x_2^n, \ldots, x_n^n\}$ of the unit ball of X and the other a subset $\{f_1^n, f_2^n, \ldots, f_n^n\}$ of the unit ball of X*, except that for a property of type P_1 the integer n is a power of 2 (an ordering can be given to the members of a set $\{x_{\varepsilon_1 \cdots \varepsilon_k}\}$ such as used for a property of type P_1 by considering $\varepsilon_1 \cdots \varepsilon_k$ to be a number written using base 3). Now let Y be the vector space consisting of all *finite* linear combinations of the infinite sequence of symbols $\{\xi_i\}$. Use a diagonal process to obtain a sequence of integers $\{k_n\}$ for which the following limits exist for all i and j and all finite sets $\{a_1, \ldots, a_r\}$ of rational numbers:

$$\lim_{n\to\infty} (f_i^{k_n}, x_j^{k_n}), \quad \lim_{n\to\infty} \| \sum_{j=1}^{r} a_j x_j^{k_n} \| .$$

Then define a norm for all finite linear combinations of members of $\{\xi_i\}$ with rational coefficients, by letting

$$\| \sum_{j=1}^{r} a_j \xi_j \| = \lim_{n\to\infty} \| \sum_{j=1}^{r} a_j x_j^{k_n} \| .$$

This norm can be extended to all of Y by the obvious limit procedure. Clearly, Y is finitely represented in X. Define a sequence of linear functionals $\{f_i\}$ on Y by letting

$$f_i(\xi_j) = \lim_{n\to\infty} (f_i^{k_n}, x_j^{k_n}) .$$

If $\xi = \Sigma_1^r \alpha_j \xi_j$, then

$$|f_i(\xi)| = |\lim_{n\to\infty} (f_i^{k_n}, \Sigma_1^r \alpha_j x_j^{k_n})| \leqq \overline{\lim_{n\to\infty}} \|f_i^{k_n}\| \, \|\Sigma_1^r \alpha_j x_j^{k_n}\|$$

$$\leqq \lim_{n\to\infty} \|\Sigma_1^r \alpha_j x_j^{k_n}\| = \|\xi\| ,$$

so that $\|f_i\| \leqq 1$. It now follows that each of (1), (2), (5), (7) or (9) that is satisfied by $\{x_k^n\}$ and $\{f_k^n\}$ for each n implies the corresponding inequality or equality for the sequences $\{\xi_k\}$ and $\{f_k\}$. Thus Y has one of the properties of type P_1^∞, P_2^∞, P_3^∞ or P_4^∞ and therefore has property ~R [see Theorems 2 and 3].

LEMMA C. *If* Y *is finitely represented in* X *and has property* ~R, *then* X *has all properties of type* P_1, P_2, P_3, P_4 *and* S.

Proof: If Y has property ~R, then it follows from Theorem 5 that Y has all properties of type P_1, P_2, P_3 and P_4. Also, it is known that Y has property S [III, Theorem 1.1, page 543]. Each of properties P_1, P_2, P_3, P_4 and S is stated for finite subspaces in such a way that X has that property if Y has the property and is finitely represented in X.

THEOREM 6. *Any two properties of type* P_1, P_2, P_3 *or* P_4 *are equivalent. Each such property is a consequence of* ~R *and each implies* S.

Proof: If a normed linear space X has property ~R, then it has all properties of type P_1, P_2, P_3 or P_4 [Theorem 5]. Also, for each property of type P_1, P_2, P_3 or P_4, that property is possessed by X if and only if it is possessed by the completion of X. Therefore, to establish the equivalence of these properties, we need consider only reflexive spaces. It follows from the three preceding lemmas that each basic property of type P_1, P_2, P_3 or P_4 implies all properties of type P_1, P_2, P_3, P_4 or S, whether or not X is reflexive. It follows from this that if X is reflexive, then each of the dual properties of type P_1, P_2, P_3 or P_4 imply all properties of type P_1, P_2, P_3, P_4 or S. Thus any two properties of type P_1, P_2, P_3 or P_4 are equivalent and each implies S.

THEOREM 7. *The property* S *is self-dual. If a normed linear space has property* S, *then it has property* $\sim U$.

Proof: To show that $S \Longrightarrow \sim U$, we recall the definitions of uniform convexity and nonsquareness and note that $U \Longrightarrow \sim S$:

"The unit ball of a normed linear space is *uniformly nonsquare* if and only if there is a positive number Δ such that there do not exist members x and y of the unit ball for which

$$\|\tfrac{1}{2}(x+y)\| > 1-\Delta \text{ and } \|\tfrac{1}{2}(x-y)\| > 1-\Delta \text{ ."} \tag{22}$$

"The unit ball of a normed linear space is *uniformly convex* if and only if, for any positive number ε, there is a positive δ such that there do not exist members x and y of the unit ball for which

$$\|\tfrac{1}{2}(x+y)\| > 1-\delta \text{ and } \|\tfrac{1}{2}(x-y)\| \geqq \varepsilon \text{ ."} \tag{23}$$

Now suppose that the unit ball is uniformly convex and choose a positive number ε and another positive number δ such that no members x and y of the unit ball satisfy (23). Let Δ be the smaller of δ and $1-\varepsilon$, so that $1-\Delta > \varepsilon$. Then no members x and y of the unit ball satisfy (22). Thus, if X is isomorphic to a space with a uniformly convex unit ball, then X is isomorphic to a space with uniformly nonsquare unit ball.

All nonreflexive Banach spaces have property S. Also, the completion of X is nonreflexive if and only if X^* is nonreflexive. Thus to show that property S is self-dual, we need consider only reflexive spaces and therefore need only show that if X has property S then X^* has property S. Suppose x and y are members of the unit ball of X for which

$$\|\tfrac{1}{2}(x+y)\| > 1-\Delta \text{ and } \|\tfrac{1}{2}(x-y)\| > 1-\Delta \text{ ,}$$

and let f and g be linear functionals with unit norms for which

$$(f, \tfrac{1}{2}[x+y]) = \|\tfrac{1}{2}(x+y)\| \text{ and } (g, \tfrac{1}{2}[x-y]) = \|\tfrac{1}{2}(x-y)\| .$$

Since neither $|f|$ nor $|g|$ is greater than 1 at either x or y, it follows that f has values greater than $1-2\Delta$ at both x and y, and g has values greater than $1-2\Delta$ at both x and $-y$. Therefore,

$$(\tfrac{1}{2}[f+g], x) > 1-2\Delta \text{ and } (\tfrac{1}{2}[f-g], x) > 1-2\Delta .$$

Thus $\|\tfrac{1}{2}(f+g)\| > 1-2\Delta$ and $\|\tfrac{1}{2}(f-g)\| > 1-2\Delta$, so that X^* has property S.

Conjectures. It is not known whether there exist Banach spaces with property S that do not have properties P_1 through P_4, or whether there exist Banach spaces with property ~U that do not have property S. It is a very interesting question whether $S \Longleftrightarrow$ ~U. For if true, then isomorphism to a space with a uniformly convex unit ball would be a self-dual property. Also, since uniform convexity of the unit ball is dual to uniform Fréchet differentiability of the norm [II, pp. 113-114], it would follow that uniform Fréchet differentiability of the norm is self-dual. A property of type P_1 was originally of interest because it seemed a natural candidate for a condition equivalent to nonisomorphism to a space with a uniformly convex unit ball. We have established one of the implications needed, that $P_1 \Longrightarrow$ ~U [see Theorems 6 and 7]. However, the following alternate proof that $P_1 \Longrightarrow$ ~U is interesting and may suggest the possibility that $S \Longrightarrow$ ~U. However, the implication ~U $\Longrightarrow P_1$ is not known to be true. It would, of course, determine the truth of all the preceding conjectures.

Direct Proof that $P_1 \Longrightarrow$ ~U: The properties of type P_1 have been shown to be equivalent to each other [Theorem 6]. Therefore, we shall start with the particular property of type P_1 which states that, for some positive number ε and any positive integer m, there is a set $\{x_{\varepsilon_1 \cdots \varepsilon_k}: 1 \leqq k \leqq m\}$ containing 2^m members of the unit ball for which each ε_i is 1 or 2 and, if $1 \leqq k < m$, then

$x_{\varepsilon_1 \cdots \varepsilon_k} = \frac{1}{2}(x_{\varepsilon_1 \cdots \varepsilon_k 1} + x_{\varepsilon_1 \cdots \varepsilon_k 2})$ and $\|x_{\varepsilon_1 \cdots \varepsilon_k 1} - x_{\varepsilon_1 \cdots \varepsilon_k 2}\| \geqq \varepsilon$.
Suppose that X has an equivalent uniformly convex $|||\ \ |||$ and that θ and ϕ are positive numbers such that

$$\theta\|x\| \leqq |||x||| \leqq \phi\|x\| \text{ for all } x .$$

Choose a positive number δ such that there do not exist x and y for which $|||x||| \leqq 1$, $|||y||| \leqq 1$,

$$|||\tfrac{1}{2}(x+y)||| > 1-\delta \text{ and } |||x-y||| \geqq \frac{\theta\varepsilon}{\phi}.$$

Then there do not exist x and y for which $|||x||| \leqq \phi$, $|||y||| \leqq \phi$,

$$(24) \qquad |||\tfrac{1}{2}(x+y)||| > \phi(1-\delta) \text{ and } |||x-y||| \geqq \theta\varepsilon .$$

Now choose m so that $\phi(1-\delta)^{m-1} < \frac{1}{2}\theta\varepsilon$. Since $|||x_{\varepsilon_1 \ldots \varepsilon_k}||| \leqq \phi$ if $k \leqq m$, and

$$|||x_{\varepsilon_1 \ldots \varepsilon_k 1} - x_{\varepsilon_1 \ldots \varepsilon_k 2}||| \geqq \theta\varepsilon \text{ if } k < m ,$$

it follows from (24) that, if $k < m$, then

$$|||x_{\varepsilon_1 \ldots \varepsilon_k}||| \leqq \phi(1-\delta)\,[\sup\{\,|||x_{\varepsilon_1 \ldots \varepsilon_k 1}|||/\phi,\ |||x_{\varepsilon_1 \ldots \varepsilon_k 2}|||/\phi]$$
$$= (1-\delta)\,[\sup\{\,|||x_{\varepsilon_1 \ldots \varepsilon_k 1}|||,\ |||x_{\varepsilon_1 \ldots \varepsilon_k 2}|||\,\}].$$

Starting with $|||x_{\varepsilon_1 \ldots \varepsilon_m}||| \leqq \phi$, m–1 repetitions of this process gives

$$|||x_\varepsilon||| \leqq (1-\delta)^{m-1} \sup\{\,|||x_{\varepsilon_1 \ldots \varepsilon_m}||| \leqq \phi(1-\delta)^{m-1} < \tfrac{1}{2}\theta\varepsilon$$

if ε is 1 or 2. This contradicts $|||x_1 - x_2||| \geqq \theta\varepsilon$.

REFERENCES

[I] E. Asplund and I. Namioka, "A geometric proof of Ryll-Nardzewski's fixed point theorem," *Bull. A. M. S., 73* (1967), 443-445.

[II] M. M. Day, *Normed Linear Spaces*, Academic Press, New York, 1962.

[III] R. C. James, "Uniformly non-square Banach spaces," *Ann. of Math.*, *80* (1964), 542-550.

[IV] ——, "Weak compactness and reflexivity," *Israel J. Math.*, *2* (1964), 101-119.

[V] J. Lindenstrauss, "On operators which attain their norms," *Israel J. Math.*, *1* (1963), 139-148.

CLAREMONT GRADUATE SCHOOL

ASYMPTOTIC FIXED POINT THEORY

by

G. STEPHEN JONES

Traditionally, topological fixed point theory has been in the large restricted to compact operators which map closed convex subsets of linear spaces into themselves. Breaking with this tradition, the notion of an asymptotically compact operator was introduced in [1] and fixed point theorems for such operators were stated which require only that closed convex sets be attractive (weakly asymptotically stable) in the orbital sense under the action of the operator. In this paper we shall modify and extend the results of [1] and [2] in the setting of a locally convex linear topological space.

X denotes a complete locally convex linear topological space. With $U \subset S \subset X$ let $f\colon S \to X$ be a mapping such that for every $j \geq 1$ the power mapping f^j is such that $f^j(U) \subset S$. Defining $f^0(U) = U$, the sequence of sets $\{f^j(U)\}$, $j = 0, 1, \ldots$, is called the *orbit* of U *under the action* of f and is denoted by $T(f, U)$. f is said to be *orbitally defined* on any set such as U. For an arbitrary set U in X, $h(U)$ denotes the closed convex hull of U. The sequence $\{(fh)^j(U)\}$, $j = 0, 1, \ldots$, where $(fh)^0(U) = h(U)$ and $(fh)^j(U) = f(h(fh)^{j-1}(U))$ for $j \geq 1$, is called the *convex orbit of* U *under the action of* f. The convex orbit of U under the action of f is denoted by $T^*(f, U)$. $T(f, U)$ (or $T^*(f, U)$) is said to be eventually contained in a set $V \subset X$ if there exists an integer $m > 0$ such that $f^j(U) \subset V$ (or $(fh)^j(U) \subset V$) for all $j \geq m$.

If a continuous operator $f\colon S \to X$ is orbitally defined on $U \subset S$ and $f^j(U)$ for some $j \geq 1$ is contained in a compact set K_0, then f is said to be *weakly eventually compact*. If $(fh)^j(U)$ for some $j \geq 1$ is contained in a compact set K_0, then fh is said to be *weakly eventually compact*.

Although not necessary for all of the results presented, we shall for convenience assume throughout that X is Hausdorff and has a countable local base. As our first example theorem we restate a simple but useful generalization of the Tychonov Fixed Point Theorem [3] first presented and proved in [2].

THEOREM 1. *Let* f *be a continuous self-mapping of a closed convex set* $U \subset X$. *If* fh *is weakly eventually compact, then* f *has a fixed point in* U.

If $f\colon S \to X$ is weakly eventually compact on $U \subset S$ and $f(S)$ is contained in a convex locally compact set, then f is said to be *eventually compact* on U. Obviously, every operator f which is compact on U is eventually compact on U. An asymptotic fixed point theorem of considerable interest from the point of view of applications may be stated as follows:

THEOREM 2. *If* $U \subset X$ *is closed and convex and* $f\colon U \to U$ *is eventually compact, then* f *has a fixed point in* U.

We shall not prove Theorem 2 since it is a special case of our Theorem 4 to be proved in the sequel.

In [4] the following important lemma was proved by Browder using the Lefschetz fixed point theorem [5].

LEMMA 1. *Let* A *be a compact absolute neighborhood retract and let* f *be a continuous self-mapping of* A. *Suppose that for some integer* m, $f^m(A)$ *is contained in a closed acyclic subset of* A. *If for some integer* m, $f^m(A)$ *is contained in a closed acyclic subset of* A, *then* f *has a fixed point.*

Another lemma which was shown to follow from Lemma 1 by Horn [6] is most convenient in proving our next theorem and may be stated as follows:

LEMMA 2. *Given a finite-dimensional complex* K, *let* K_0 *be a closed, bounded, and acyclic subcomplex. If* f *is a simplicial mapping of the* k^{th} *barycentric subdivision of* K *into* K *such that, for some integer* m, $f^j(K_0) \subset K_0$ *for* $m \leq j \leq 2m$, *then* f *has a fixed point in* K_0.

Our next result is an extension of Theorem 3 in [2] and a stronger theorem for a general Banach space proved by Horn in [7]. It can also be considered as a modification and extension to more general spaces of a theorem of F. Browder in [4] for Banach spaces and D. Bourgin for the finite-dimensional case.

THEOREM 3. *Let* $V \subset U \subset Y$ *be subsets of* X *with* V *and* Y *compact and open relative to* Y. *Let* $f\colon Y \to X$ *be continuous and such that* $T(f, U) \subset Y$. *If* $T(f, U)$ *is eventually contained in* V, *then* f *has a fixed point in* V.*

Proof: This theorem for the case where X is a Banach space and in particular a finite-dimensional Euclidean space is presented in [7] and proved through the use of Lemma 2. In order to shorten our presentation here, we will, therefore, assume the result for the finite-dimensional case.

Since X is Hausdorff and has a countable local base it is metrizable. Let ρ be a metric yielding the topology of X. Since Y is closed and convex it follows from a theorem of Dugundji in [8] that there exists a retraction $r\colon X \to Y$. Defining $F = rf$ we have that $F(Y) \subset Y$ and our hypotheses clearly imply the existence of an integer $m > 0$ such that $F^j(U) \subset Y$ for all $j \geq m$. Clearly any fixed point of F in V is also a fixed point of f.

---

* If instead of assuming U is open relative to Y we assume U is internal to Y in the sense of convexity, then the theorem thus modified may be proved in essentially the same manner used to prove Theorem 3 in [2] by direct use of Browder's Theorem 2 in [4].

Let $G(\eta)$, $\eta > 0$, denote the set of all continuous maps $g\colon Y \to Y$ such that $\rho(g(x), F(x)) < \eta$ for all $x \in Y$. The continuity of F and the compactness of Y imply that for every $\varepsilon > 0$ there exists $\eta > 0$ such that $g \in G(\eta)$ implies $\rho(g^j(x), F^j(x)) < \varepsilon$ for $x \in Y$ and $1 \leq j \leq 2m$. Let $\{\varepsilon_i\}$, $i \geq 1$, be any positive sequence tending to 0 and let $\{\eta_i\}$ be such that $g \in G(\eta_i)$ implies $\rho(g^j(x), F^j(x)) < \varepsilon_i/2$ for all $x \in Y$ and $1 \leq j \leq 2m$. Let Z_i be a finite subset of Y such that $Z_i \cap V$ is nonvoid and $y \in Y$ implies $\rho(y, z) < \eta_i/5$. Let L_i denote the linear extension of Z_i, $Y_i = Y \cap L_i$, $U_i = U \cap L_i$. and $V_i = V \cap L_i$. There exists a triangulation $T_i\colon K_i \to Y_i$ for some finite-dimensional complex K_i. Furthermore, there exist $\delta_i > 0$ such that $\rho(x, y) < \delta_i$ implies $\rho(F(x), F(y)) < \eta_i/5$ and we may assume the mesh of T_i is less then δ_i. For each point $\omega = T_i(v)$ in Y_i where v is a vertex of K_i, let $g_i(\omega)$ be some $z \in Z_i$ such that $\rho(z, F(\omega)) < \eta_i/5$. $g_i^* = T_i^{-1} g_i T_i$ is a mapping defined on the vertices of K_i and we extend this mapping linearly on all simplexes of K_i. We extend g_i to a mapping on all of Y_i into itself by the formula $g_i = T_i g_i^* T_i^{-1}$. It is easily verified that $\rho(F(x), g_i(x)) < \eta_i$ for all $x \in Y_i$. Since $F^j(U) \subset V$ for $m \leq j \leq 2m$, it follows that

$$g_i^j(U_i) \subset (V + N_{e_i}) \cap Y_i = V_i^*$$

for $m \leq j \leq 2m$ where N_{ε_i} is a ρ-neighborhood of the origin with radius ε_i. Now $V_i \subset U_i$ and $g_i^j(U_i) \subset V_i$ for $j \geq m$, so it follows from the validity of our theorem in the finite-dimensional case that g_i has a fixed point x_i in V_i for all i. But, of course, the sequence of fixed points $\{x_i\}$ is contained in the compact set Y and therefore, has a limit point. Clearly any such limit point is a fixed point of F and consequently f and is contained in V. Thus our proof is complete.

As an extension of Theorem 3 we have the following result.

THEOREM 4. *Let* $S \subset X$ *be closed and convex and let* $f\colon S \to X$ *be eventually compact on* $U \subset S$ *where* U *is convex and open relative to* S.

If $T(f, U)$ *is eventually contained in a closed convex set* $V \subset U$, *then* f *has a fixed point in* V.

Proof: Since f is eventually compact on U there exists a positive integer m_1 such that for each $j \geq m_1$, $f^j(U)$ is contained in a compact set. Our hypotheses also imply that there exists a positive integer m_2 such that for all $j \geq m_2$, $f^j(U) \subset V$. Hence if $m = m_1 + m_2$ we have that $V_0 = f^m(U)$ is a conditionally compact subset of V. Furthermore, $V_1 = h(\bigcup_{i=0}^{m-1} f^i(V_0))$ is a compact convex subset of V and $f^j(U) \subset V_1$ for all $j \geq m$.

Now again by the eventual compactness of f there exists a convex locally compact set $Q \subset X$ such that $f(S) \subset Q$. There exists an open neighborhood N of the origin such that $V_1 + N \cap Q$ is conditionally compact and $V_1 + N \cap S \subset U$. Defining $U_1 = V_1 + N \cap Q \cap S$ we clearly have that U_1 is conditionally compact, open relative to $Q \cap S$, and $U_1 \subset U$. Defining $Y_1 = h(\bigcup_{i=0}^{m-1} f^i(U_1))$ we have that $Y_1 \subset Q \cap S$ and Y_1 is convex and compact and U_1 is open relative to Y_1. f with respect to the sets $V_1 \subset U_1 \subset Y_1$ satisfies the hypotheses of Theorem 3. Hence we may conclude from Theorem 3 that f has a fixed point in V_1 and our proof is complete.

We note at this point that Theorem 2 is a special case of Theorem 4.

Let Γ be a directed set and consider a net of continuous functions $f_\gamma \colon S \to X$, $\gamma \in \Gamma$. The notation $\{f_\gamma\} \to f$ specifies that for an arbitrary neighborhood N of the origin there exists $\gamma^* \in \Gamma$ such that for all $x \in S$,

$$f_\gamma(x) \in f(x) + N,$$

whenever $\gamma \geq \gamma^*$. $\{f_\gamma\}$ is called a *generator* for f if $\{f_\gamma\} \to f$. $\{f_\gamma\}$ is called a *regular generator* of f if f_γ, having a fixed point for each γ sufficiently large, implies f has a fixed point.

If $\{f_\gamma\}$ is a regular generator for f and each mapping f_γ is weakly eventually compact on $U \subset S$, then f is said to be *weakly asymptotically compact* on U. Likewise if each $f_\gamma h$ is weakly eventually compact on $U \subset S$, then fh is said to be *weakly asymptotically compact.* An immediate extension of Theorem 1 for weakly asymptotically compact operators may be stated as follows.

THEOREM 5. *If* U *is a closed convex set in* X *and* f: $U \to U$ *is such that* fh *is weakly asymptotically compact, then* f *has a fixed point in* U.

If $\{f_\gamma\}$ is a regular generator for f and each mapping f_γ is eventually compact on $U \subset S$, then f is said to be *asymptotically compact on* U. If there exists a compact convex set K such that for every $\gamma \in \Gamma$, $T(f_\gamma, U)$ is eventually contained in K, then $\{f_\gamma\}$ is said to be *eventually equicompact.* If f has an eventually equicompact generator on U, then clearly f is asymptotically compact on U.

THEOREM 6. *If* $U \subset X$ *is closed and convex and* f: $U \to U$ *is asymptotically compact, then* f *has a fixed point in* U.

Theorem 6 follows immediately from Theorem 2 and the definition of asymptotic compactness.

We note in passing that the asymptotically compact mapping forms a class significantly more general than the compact mappings. As a consequence, a fixed point theory for such mappings is more broadly applicable in the theory of functional equations; in particular to partial differential equations and to hereditary differential equations such as discussed in [8]. Projectionally compact mappings as introduced by Petryshyn in [9] and [10] are a special subclass of asymptotically compact mappings. Hence the rather extensive connections in the literature pointed out in these papers are contained within the theory developed in [1] and continued here. We also point out that the fixed point theory derived through the use of the homotopy extension theorem carries over with little difficulty to asymptotically compact mappings. Some developments in this direction are discussed in [11].

THEOREM 7. *Let* $S \subset X$ *be closed and convex, and let* f: $S \to X$ *be approximated by the generator* $\{f_\gamma\}$, $\gamma \in \Gamma$, *and be asymptotically compact on*

$U \subset S$ where U is convex and open relative to S. If for each $\gamma \in \Gamma$, $T(f_\gamma, U)$ is eventually contained in a closed convex set $V_\gamma \subset U$, then f has a fixed point in U.

Proof: Each mapping f_γ, $\gamma \in \Gamma$, has a fixed point in V_γ by Theorem 4. Hence from the definition of a regular generator the mapping f has a fixed point in U.

Consider a function $g: S \to X$ and a set $V \subset S$. Suppose there exists a neighborhood M of the origin such that f is orbitally defined on $(V+M) \cap S$ and for each neighborhood N of D there exists n_1 such that

$$g^j((V+M) \cap S) \subset (V+N) \cap S$$

for all $j \geq n_1$. Borrowing terminology from the theory of dynamical systems we refer to V as an *attractor* under the action of g. For mappings with attractors we have the following theorem.

THEOREM 8. *Let $S \subset X$ be closed and convex, and let $f: S \to X$ be approximated by the generator $\{f_\gamma\}$, $\gamma \in \Gamma$, and be asymptotically compact on $U \subset S$ where U is convex and open relative to S. If $V \subset U$ is a closed convex attractor under the action of f_γ for all γ sufficiently large, then f has a fixed point in V.*

In this paper we have, for the most part, been examining a fixed point theory from the point of view of stability of sets generated by the powers of an operator. Such a point of view naturally suggests the possibility for developing a Liapunov type theory for fixed points. It is, in fact, possible to do so even in the context of very general linear spaces. A result of particular interest along these lines is an analogue of LaSalle's Theorem [12] which is developed for determining the existence of fixed points in [13].

REFERENCES

[1] Jones, G. S., "Stability and asymptotic fixed point theory," *Proc. Nat. Acad. Sci., U.S.A., 53* (1965), 1262-1264.

[2] _____, "Asymptotic fixed point theorems and periodic systems of functional-differential equations," *Contrib. Differential Eqs., 2* (1963), 385-405.

[3] Tychonov, A., "Ein Fixpunktstaz," *Math. Ann., 111* (1935), 767-776.

[4] Browder, F. E., "On a generalization of the Schauder fixed point theorem," *Duke Math. J., 26* (1959), 291-303.

[5] Lefschetz, S., "Algebraic Topology," *Amer. Math. Soc. Colloquium Pub., 27* (1942).

[6] Horn, W. A., "A generalization of Browder's fixed point theorem," *Not. Amer. Math Soc.,* Abstract No. 611-40, 11, 325, April (1964).

[7] _____, *Fixed Point Theorems for Iterated Mappings,* Ph.D. dissertation, University of Maryland (1967).

[8] Dugundji, J., "An extension of Tietze's theorem," *Pac. J. Math., 1* (1951), 353-367.

[9] Petryshyn, W. V., "On a fixed point theorem for nonlinear P-compact operators in Banach space," *Bull. Amer. Math. Soc., 72* (1966), 329-334.

[10] _____, "Further remarks on nonlinear P-compact operators in Banach space," *Proc. Nat. Acad. Sci.,* U.S.A., *55* (1966), 684-687.

[11] Jones, G. S., "Restrictive resolvable operators and fixed point theory," University of Maryland report BN-500; *Mathematical Systems Theory* (1967).

[12] LaSalle, J.P., "Some extensions of Liapunov's second method," IRE Trans. *Prof. Group on Circuit Theory,* CT-7 (1960), 520-527.

[13] Jones, G. S., "Fixed Point Theorems for Asymptotically Compact Mappings," *University of Maryland Report* BN-503 (1967).

THE UNIVERSITY OF MARYLAND

ON THE RELATIONSHIP OF LOCAL TO GLOBAL FIXED-POINT INDEXES

by

RONALD J. KNILL

For the category of open subsets of Euclidean space and continuous maps, the Lefschetz index assigns an integer $\Lambda(f)$ to each map $f\colon X \to X$, where the closure of $f(X)$ is a compact subset of X. This integer is a *fixed-point* index since if $\Lambda(f) = 0$, then $f(x) = x$ for some $x \in X$. Such an index is a *global index* since it is defined only when $f(X) \subset X$. On the other hand, Leray's adaptation of Brouwer degree to a fixed point index is an index which assigns an integer $i(f)$ to each map $f\colon X \to Y$, where X is an open subspace of Y, the closure of $f(X)$ is compact in Y, and f has a continuous extension to the closure of X in Y such that the extension has the same fixed-point set, F, as f. This is a *local index* since if C is a connected component of F, then $i(f \mid V)$ is defined and constant on all sufficiently small neighborhoods of C.

In order to further discriminate between local and global indexes it is necessary to define what is meant by a (fixed-point) index on a category $\underline{T}$ of topological spaces and continuous maps. Such a definition will be given in terms of properties well known to be common to the Lefschetz and Leray indexes, but which allows indexes to take their values in principal ideal domains. After a brief discussion of known results, we will exhibit a category $\underline{T}$ of topological spaces and maps, which contains ANR spaces and quasi-complexes, which has arbitrary products, and on which the Lefschetz index with rational values is a fixed-point index. The Lefschetz index is defined in terms of Čech cohomology with rational coefficients.

It turns out that Cech homology with integer coefficients is inadequate in this context, for if the Lefschetz index were to be defined in terms of this homology functor, it would not be a *fixed-point* index on $\underline{T}$. Also it is not possible to define a local fixed-point index on $\underline{T}$ which extends the Leray index. Detailed definitions and statements of theorems are provided, but proofs will appear in full elsewhere. We conclude with a short list of problems.

§1. *Indexes and Fixed-point Indexes.*

All spaces are regular Hausdorff spaces. Suppose f: $X \to Y$ is a map such that $X \subset Y$. *A fixed point* of f is a point x of X such that $f(x) = x$. A *virtual* fixed point of f is a point x of the closure, X^{-}, of X such that x is in the adherence of the filter base of sets of the form $f(V \cap X)$, where V ranges over neighborhoods of x in Y. Let $\underline{T}$ be a category of regular Hausdorff spaces for objects and of continuous maps for morphisms. Let $\underline{J}(\underline{T})$ be the class of all maps f: $X \to Y$ of $\underline{T}$ such that X is an open subspace of Y, f(X) is relatively compact in Y, and such that f has no virtual fixed points on the boundary of X.

1.1 *Definition.* A *fixed-point index* on $\underline{T}$ is a pair $(i, \underline{J})$ such that $\underline{J} \subset \underline{J}(\underline{T})$ and i is a function which assigns to each $f \in \underline{J}$ an element i(f) of a principal ideal domain, P, called the *coefficient ring* of the index, subject to the following conditions on i and $\underline{J}$.

1.1.1 If f: $X_1 \to Y_1$ and g: $X_2 \to Y_2$ are maps of $\underline{J}$, then the maps

$$f \times g: X_1 \times X_2 \to Y_1 \times Y_2 \quad [1]$$

[1] $X_1 \times X_2$ and $Y_1 \times Y_2$ denote the topological (Tychonoff) product spaces, respectively, and $(f \times g)(x_1, x_2) = (fx_1, gx_2)$, for $(x_1, x_2) \in X_1 \times X_2$.

and

$$f + g\colon\ X_1 + X_2 \to Y_1 \times Y_2$$ [2]

are in $\underline{J}$ and

$$i(f \times g) = i(f) \times i(g)$$

$$i(f + g) = i(f) + i(g) \ .$$

1.1.2. Let $[0, 1]$ be the closed unit interval. If

$$h\colon\ X \times [0, 1] \to Y$$

is a homotopy such that for each $t \in [0, 1]$, $h_t \in \underline{J}$, and such that the image, $h(X \times [0, 1])$, is relatively compact in Y, then

$$i(h_0) = i(h_1) \ .$$

1.1.3. If $f\colon X \to X$ is a constant map and $f \in \underline{J}$, then

$$i(f) = 1 = \text{the identity of } P.$$

1.1.4. Suppose $f\colon X \to Y$ factorizes as $f = uf'$, where $f'\colon X \to Y'$ has relatively compact image in Y', and f' and u are continuous, and let $X' = u^{-1}(X)$, and $u' = u|X'$. Then $f'u'\colon X' \to Y'$ is in $\underline{J}$ if f is in $\underline{J}$ and

$$i(f'u') = i(f) \ .$$

1.1.5. (Fixed-point property). If $f\colon X \to Y$ is in $\underline{J}$ and X' is an open neighborhood in X of the set, F_f, of fixed points of f, and if $f|X'$ is in $\underline{J}$, then

$$i(f|X') = i(f) \ .$$

---

[2] $X_1 + X_2$ and $Y_1 + Y_2$ denote the disjoint topological sums of X_1, X_2 and of Y_1, Y_2, respectively. The map $f + g$ is defined by

$$(f+g)(x) = \begin{cases} fx & \text{if } x \in X_1 \\ g(x) & \text{if } x \in X_2 \end{cases} \ .$$

This last property is called the *fixed-point property,* for with it, with properties 1.1.1 and 1.1.4, and with the assumption that the empty map $\emptyset\colon \emptyset \to \emptyset$ is in $\underline{J}$, one may show that for a map f in $\underline{J}$ which has no fixed points, $i(f) = 0$. It is the case that $(i, \underline{J})$ can satisfy 1.1.1 through 1.1.4 without satisfying 1.1.5. In such a case call $(i, \underline{J})$ an *index.* A *local index* on $\underline{T}$ is an index $(i, \underline{J})$ such that if $f\colon X \to Y$ is a map of $\underline{J}$ and C is a connected component of the fixed point set of f, and if U is an open neighborhood of C such that the only fixed points of f in $\overline{U}$ are in C, then $f|U$ is in $\underline{J}$, and $i(f|U)$ is independent of choice of such a U.

Classical Case, Part 1. (Hopf-Lefschetz). Let E be the category of triangulable compact spaces and continuous maps. Let $\underline{J}$ be the class of maps $f\colon X \to X$ of $\underline{E}$, and let $\Lambda(f)$ be the Lefschetz number of f:

$$\Lambda(f) = \sum_{n \geq 0} \operatorname{tr} H_n(f; Q)$$

where $H_n(f; Q)$ is the simplicial homology homomorphism, rational coefficients, which is induced by f, and "tr" = "trace of." In this case $(\Lambda, \underline{J})$ is a fixed-point index of $\underline{E}$, with values in the ring of integers.

Several authors, most prominently Lefschetz, have extended this result to larger categories of spaces [13], [14], [15], [17], and [1]. Browder [6] and Eells [7] have pointed out that the Hopf-Lefschetz theorem for the category $\underline{T}$ of topologically complete ANR (metric) (Hanner's terminology [8]) spaces follows from the corresponding theorem for the category of open subspaces of Banach spaces and continuous maps. In the latter case one may use singular or Čech homology in place of simplicial homology. Actually these authors were concerned with showing that a map $f\colon X \to X$ such that $f(X)^-$ is compact and $\Lambda(f) \neq 0$, has a fixed point; however, the broadening of the definition of $\Lambda(f)$ to maps $f\colon X \to Y$ such that X is open in Y, Y is an ANR (metric) space, $f(X)^-$ is compact and a subset of X, presents no real difficulties. We will prove a slightly more extensive theorem shortly.

Leray [15], [16], [18] noted that it was possible to define a fixed-point index i(f) for a map f: X → Y such that f has no virtual fixed points on the boundary of X, rather than $f(X)^{-} \subset X$. This index is related in a natural way to Schauder-Leray degree.

Classical Case, Part 2. (Leray) Let $\underline{C}$ be the category of open subspaces of convexoide compact spaces and continuous maps. There is a fixed-point index (i, $\underline{J}(\underline{C})$) on $\underline{C}$ with coefficient ring the integers and such that whenever f: X → X is in $\underline{J}(\underline{C})$ then i(f) = Λ(f).

In the case of the Leray index, properties 1 – 4 of the index are useful in computing the index of a map, and as O'Neill [19] showed, slightly stronger versions of these properties characterize the Leray index on $\underline{E}$, provided it agrees with the Lefschetz index whenever both indexes are defined on a map of $\underline{E}$. The Leray index also provides a device for investigating essential sets of fixed points [10], [19], and [4].

Deleanu [22] and Browder [5] have extended Leray's results. The following theorem, whose formulation is suggested by the Browder-Eells result, has a proof based on a method which is well-known, namely reduce the theorem to one about convex subsets of locally convex topological vector spaces.

1.2 THEOREM. *If* N *is the category of neighborhood extensors for compact spaces, then there is a fixed-point index* (i, $\underline{J}(\underline{N})$) *on* $\underline{N}$ *with the integers for coefficient ring and such that* i(f) = Λ(f), *whenever the right-hand side is defined for a map* f *of* $\underline{J}(\underline{N})$.

Proof: Suppose f: X → Y is a map in $\underline{J}(\underline{N})$ and that A is the B*-algebra of continuous complex-valued maps on $f(X)^{-}$. Let A* be the space of continuous linear functionals on A, and equip A* with the weak-star topology. Then there is a homeomorphic image of $f(X)^{-}$ in A* and a finite union, Y′, of compact convex subsets of A* which contain this image. Since Y is a neighborhood extensor for compact spaces, then we may assume that these compact convex sets are chosen in a small enough neighborhood of $f(X)^-$, so that there is a map $u\colon Y' \to Y$ such that if $v\colon f(X)^- \to Y'$ is the natural imbedding, then $u \circ v$ is the inclusion of $f(X)^-$ into Y. Let

$$f' = v \circ f\colon X \to Y'$$

$$X' = u^{-1}(X)$$

$$u' = u \,|\, X' \;.$$

Now Y' is convexoide, so Leray's theorem applies and $i(f' \circ u')$ is defined. Let

$$i(f) = i(f' \circ u')$$

Then the properties 1.1.1 through 1.1.5 of the Leray index imply the corresponding properties for $(i, \underline{J}(\underline{N}))$. One could provide an alternative proof using methods analogous to those of Granas [23] and Klee [24].

Note that any ANR (normal) or ANR (metric) space is a neighborhood extensor for compact spaces, and so is any C^0-manifold modelled on a metrizable locally convex topological vector space.

§2. *The Significance of the Coefficient Ring.*

In this section it will be shown by an example that the field of rationals is preferred over the integers as the coefficient ring for the Lefschetz fixed-point index and that Čech cohomology is preferable to homology for defining the Lefschetz index. In order to give the relevant definitions it will be necessary to introduce a modification of Lefschetz's quasi-complexes: "P-simplicial spaces," where P is a principal ideal domain. The essential idea of our modification is to eliminate Lefschetz's conditions involving chain homotopies.

Let P be a principal ideal domain and let Q be its field of rationals. P is canonically imbedded into Q. Let $\Phi\colon W \to W$ be an endomorphism of a P-module, W, and let $N(\Phi)$ be the submodule of W formed by those elements w such that some iterate, $\Phi^n(w)$, is zero, $n \geq 1$. We say that Φ has *finite type* if $W/N(\Phi)$, modulo its torsion subgroup, is a free P-module of finite rank. In such a case, Leray defines the trace $tr(\Phi)$ of Φ to be the trace of the natural automorphism induced by Φ on $Q \otimes W/N(\Phi)$. Let Y be a regular Hausdorff space.

A map $f\colon X \to Y$ is said to be P-*admissible* if X is an open subspace of Y, $f(X)^-$ is a compact subset of X, and the corestriction to X

$$f_X\colon X \to X$$

induces a Čech cohomology homomorphism

$$H^*(f_X;\, P)\colon\ H^*(X;\, P) \to H^*(X;\, P)$$

which is of finite type. In this case, let

$$\Lambda_P(f) = \sum_{n \geq 0} (-1)^n\, tr(H^n(f_X;\, P))\, .$$

If $\Lambda_P(f)$ exists, so does $\Lambda_Q(f)$ and they are equal, according to the universal coefficient theorem, which applies here since $f(X)^-$ was assumed to be compact. If $\underline{T}$ is a category of regular Hausdorff spaces and continuous functions, let $\underline{J}_P(\underline{T})$ be those P-admissible maps $f\colon X \to Y$ of $\underline{T}$, such that for some compact subset A of Y which contains the image of f, the map $f_A\colon X \to A$, which is the corestriction of f, induces a Čech cohomology homomorphism $H^*(f;\, Q)$, such that $H^*(f;\, Q)$ has an image which is a finite dimensional vector space over Q. Then $(\Lambda_P, \underline{J}_P(T))$ is the *Lefschetz index* on $\underline{T}$. In general, it is not a fixed point index, as one can conclude from Borsuk's classic example [2], or the simpler example of Wilder [20].

In order to define a reasonably inclusive category $\underline{T}$ for which $(\Lambda_P, \underline{J}_P(\underline{T}))$ is a fixed-point index, we make use of Vietoris chains. Let Y be a space, and let α be a family of open subsets of Y. For a subset X of Y, let $X(\alpha) = \{X_n(\alpha)\}_{n \geq 0}$ be the simplicial set such that $X_n(\alpha)$, $n \geq 0$, consists of those $n+1$-tuples $s = (s_0, s_1, \ldots, s_n)$ of elements of X such that if $|s| = \{s_i\colon i = 0, 1, \ldots, n\}$, then $|s| \subset A$ for some $A \in \alpha$. The ith face operator d_i on $X_n(\alpha)$ takes s into $(s_0, s_1, \ldots, s_{i-1}, s_{i+1}, \ldots, s_n) \in X_{n-1}(\alpha)$. The ith degeneracy operator is also the standard one, but we shall not need it here. Let $C_*(X(\alpha); P)$ be the free chain complex on $X_n(\alpha)$ with coefficients in P. If

$$\partial c = p_1 s^1 + p_2 s^2 + \cdots + p_n s^k \in C_n(X(\alpha); P) ,$$

then $|c| = \bigcup\{s^j\colon p_j \neq 0\}$. The boundary ∂c is

$$\partial c = \sum_{j=1}^{k} p_j \sum_{i=0}^{n} (-1)^i d_i(s^j) .$$

2.1. *Definition.* With X and Y as above and with α, β as families of open subsets of Y, a chain map

$$\Phi\colon C_*(X(\alpha); P) \to C_*(Y(\beta); P)$$

is a P-module homomorphism which preserves degree and which commutes with ∂. Φ is said to be *carried* by a subset U of $Y \times Y$ if for $s \in X(\alpha)$ and $p \in P$,

$$|s| \times |\Phi(ps)| \subset U .$$

2.2. *Definition.* A compact subset X of a regular Hausdorff space Y is said to be P-*simplicially* imbedded (in Y) if for any open neighborhood U in $Y \times Y$ of the diagonal, $\{(x, x)\colon x \in X\}$, of X, there is an open covering α of X such that for any open covering β of Y, there exists a chain map carried by U

$$\Phi\colon C_*(X(\alpha); P) \to C_*(Y(\beta); P) .$$

Y is said to be P-*simplicial* if every compact subset of Y is P-simplicially imbedded in Y.

2.3. THEOREM. *The following spaces are* P-*simplicial:*

2.3.1. *Quasi-complexes;*

2.3.2. *A space* Y *of covering dimension* $\leq n$, *and which is* LC^{n-1}, n *finite* (see [25] for the definition of LC^{n-1});

2.3.3. *A cone over a* P-*simplicial space.*

2.3.4. *An arbitrary product of* P-*simplicial spaces;*

2.3.5. HLC* *spaces.*

2.4. THEOREM. *Suppose that* G *is a compact totally disconnected group such that for every open subgroup* G, *elements of* P *are divisible by the order of* G/H. *Let* X *be a compact space on which* G *acts freely and such that the orbit space* X/H *is* P-*simplicial for each open subgroup,* H, *of* G. *Then* X *is* P-*simplicial.*

2.5. COROLLARY. *If every element of* P *is divisible by an integer* $q \geq 2$, *then the* q-*adic solenoid is* P-*simplicial.*

2.6. THEOREM. *If* $\underline{T}$ *is the category of* P-*simplicial spaces and continuous maps, then* $(\Lambda_P, \underline{J}_P(\underline{T}))$ *is a fixed-point index of* $\underline{T}$. *Furthermore, there is no local fixed-point index* $(i, \underline{J})$ *on* $\underline{T}$ *for which* $i \mid \underline{J}_P(\underline{T}) = \Lambda_P$.

Example: Let X be the q-adic solenoid and let f: $X \to X$ be the qth root map. Then $\Lambda_Q(f) = 1 - 1/q$ (if $Q = Q(Z)$, Z = integers) so $\Lambda_Q(f) \notin Z$. On the other hand, the Čech homology of X is that of a point, so that if one defines $\Lambda_Z(f)$ in terms of the Čech homology homomorphism induced by f, then $\Lambda_Z(f) = 1$ for any f: $X \to X$, despite the fact that f could be a translation and would have no fixed points in this case. Thus Čech homology is unsuitable for defining Λ_Z.

Comments. Combining 2.1.2 with 2.6 shows that the Hopf-Lefschetz theorems can be extended to the category of paracompact Hausdorff spaces of covering dimension $\leq n$ and which are LC^{n-1} at every point. This improves a result of ours [12], and as a consequence of an example of Bing [21], the condition "LC^{n-1}" cannot be relaxed to "LC^{n-2}".

Parts 2.3.1 and 2.3.4 of Theorem 2.3 provide an improvement of a result of Brahana [3] that states that any *finite* product of quasicomplexes is a quasicomplex. Our paper [12] is of relevance to this also.

It may seem strange to define P-simplicial in terms of chains when Λ_P is defined in terms of cohomology, and in fact one could easily translate this definition into one in terms of Alexander cohomology as for example, we reported in [11]. However, the chain formulation is easier to handle in regard to products, since it permits the use of the method of acyclic carriers.

It is possible to make a more precise distinction between categories $\underline{T}$ on which $(\Lambda_P, \underline{J}_P(\underline{T}))$ is a fixed-point index, and categories on which there is a local fixed-point index. One does so with a functor which is defined by extending the methods of Leray [17]. Without specifically defining this functor we may express the formalism involved. Suppose that $\underline{M}$ is the category whose objects are the inclusion maps $f\colon X \to Y$ of compact subspaces X of a regular Hausdorff space Y. A morphism of $\underline{M}$ is a pair of maps (u, v),

$$(u, v)\colon\ f \to g$$

such that $f = u \circ g \circ v$.

2.7. THEOREM. *For a principal ideal domain,* P, *there is a functor* $L\colon \underline{M} \to \underline{A}$, *where* A *is the category of* P-*module homomorphisms, such that:*

2.7.1. *If* $\underline{T}$ *is the category whose spaces are all regular Hausdorff spaces and whose maps are identities, or maps* $f\colon X \to Y$ *such that* X *is open in* Y, $f(X)^-$ *is compact in* X *and such that if* $j\colon f(X)^- \to Y$ *is inclusion, and* $L(j)$ *is a monomorphism, then* $(\Lambda_P, \underline{J}_P(\underline{T}))$ *is a fixed-point index on* $\underline{T}$.

2.7.2. *If* $\underline{T}$ *is the category whose objects are all regular Hausdorff spaces and whose maps are identities, or maps* $f\colon X \to Y$ *such that* X *is open in* Y, $f(X)^-$ *is compact in* Y *and such that if* $j\colon f(X)^- \to Y$ *is inclusion, and* $L(j)$ *is an isomorphism, then there exists a local fixed-point index on* $\underline{T}$.

§3. *Problems.*

One of the simplest and most useful methods of characterizing "nice spaces" is by local properties. This has been done among paracompact spaces for such classes as metric, complete metric, ANR(Q) (where Q is any of several classes) spaces. It would be useful also to be able to define a category $\underline{T}$ on which there is a local fixed-point index, according to simple local properties of its object spaces.

One such property is the property of local contractibility. Specifically:

Q. Let $\underline{T}$ be the category of all locally contractible regular Hausdorff spaces. Is there a local fixed-point index on $\underline{T}$?

Our result 2.5 together with 2.6 shows that the Lefschetz index, (rational coefficients) is a fixed-point index on the category of solenoids. It is well known that it is a fixed-point index on the category of Lie groups and continuous functions, since such groups are ANR (metric) space.

Q. Is the Lefschetz index with rational coefficients a fixed-point index on the category of all topological groups and continuous functions?

There is a well known problem common to both these questions, namely:

Q. Let X be an open convex subset of a topological vector space over the real field. Let f: $X \to X$ be a map such that the closure of f(X) is a compact subset of X. Does f have a fixed point?

REFERENCES

[1] Begle, E., "A fixed-point theorem," *Ann. of Math., 51* (1950), 544-550.

[2] Borsuk, K., "Sur un continu acyclique qui se laisse transformer topologiquement et lui meme sans points invariants," *Fund. Math., 24* (1935), 51-58.

[3] Brahana, Thomas R., "Products of Quasi-Complexes," *Proc. Amer. Math. Soc., 7* (1956), 954-958.

[4] Browder, F., "On the continuity of fixed points under deformations of continuous mappings," *Summa Brazil. Math., 4* (1960), 183-191.

[5] Browder, F., "On the fixed-point index for continuous mappings of locally connected spaces," *Summa Brazil. Math.*, *4* (1960), 253-293.

[6] _____, "Fixed-point theorems on infinite dimensional manifolds," *Trans. Amer. Math. Soc.*, *119* (1965), 179-194.

[7] Eells, J., "A setting for global analysis," *Bull. Amer. Math. Soc.*, *72* (1966), 797.

[8] Hanner, O., "Retraction and extension of mappings of metric and nonmetric spaces," *Ark. Mat.*, *2* (1952), 315-360.

[9] Hu, S., *Theory of Retracts,* Wayne State University Press, Detroit, 1965, p. 28.

[10] Kinoshita, S., "On essential components of the set of fixed points," *Osaka Math. J. 4,* 19-22.

[11] Knill, R., "On the Lefschetz fixed-point theorem," (abstract) *Amer. Math. Soc. Notices, 11* (1964), 614-648.

[12] _____, "Cones, products, and fixed points," *Fund. Math. LX* (1967), 35-46.

[13] Lefschetz, S., "On locally connected and related sets," *Ann. of Math.*, *35* (ser. 2) (1934), 118-129.

[14] _____, "Algebraic Topology," *Amer. Math. Soc. Colloquium Pub.*, *27* (Chapter VIII), New York, 1942.

[15] Leray, J., "Les equations dans les espaces topologiques," *C. R. Acad. Sci. Paris, 214* (1942), 897-899.

[16] _____, "Transformations et homeomorphies dans les espaces topologiques," *C. R. Acad. Sci. Paris, 214* (1942), 938-940.

[17] _____, "Sur les equations et les transformations," *J. Math. Pures Appl.*, *24* (1946), 201-248.

[18] _____, "Theorie des points fixes: indice total et nombre de Lefschetz," *Bull. Soc. Math. France, 87* (1959), 221-233.

[19] O'Neill, B., "Essential sets and fixed points," *Amer. J. Math.*, *75* (1953), 497-509.

[20] Wilder, R., "Some mapping theorems with applications to nonlocally connected spaces," *Algebraic Geometry and Topology*, a symposium in honor of S. Lefschetz, Princeton University Press, 1957, Princeton, New Jersey.

[21] Bing, R., "Challenging Conjectures," *Amer. Math. Monthly*, *74* (1967), 56-64.

[22] Deleanu, S., "Theorie des points fixes: sur les retracts voisinages des espaces convexoïdes," *Bull. Soc. Math. de Fr.*, *87* (1959), 235-243.

[23] Granas, A., *Introduction to Topology of Functional Spaces*, Lectures [Chicago] 1961.

[24] Klee, V., "Leray-Schauder theory without local convexity," *Math. Annalen*, *141* (1960), 286-296.

[25] Borsuk, K., *Theory of Retracts*, Warsaw, 1967.

TULANE UNIVERSITY

C^1-EQUIVALENCE OF FUNCTIONS NEAR ISOLATED CRITICAL POINTS

Nicolaas H. Kuiper

Summary. We study local properties of functions on R^n or Hilbertspace near an isolated critical point. In Theorems 1 and 2, which are formulated in §1 and proved in §4, we obtain local C^0- or C^1-equivalence of functions under conditions in which a property $Q(r)$ plays a key rôle. In §2 we discuss this property, which generalizes nondegeneracy $= Q(2)$. In §3 we give some examples which restrict the scope of possible generalizations. Finally in the appendix we present in a short formulation Palais' proof of the Morse lemma.

Unfortunately we did not succeed in getting conclusions involving higher differentiability (C^s-equivalence $s > 1$) analogous to those we obtained in [1] for nondegenerate critical points.

§1. C^s-*sufficiency; formulation of the main theorems*

Let V be either n-dimensional euclidean space R^n or real Hilbert-space H. We consider functions $f: V \to R$ of differentiability class $t \leq \infty$ (C^t-functions) and we are in particular interested in their *local behaviour* near the point $0 \in V$. We assume $f(0) = 0$ and let $\infty + 1 = \infty$. Following Ehresmann we say that two C^t-functions $f, g: (V, 0) \to (R, 0)$ are r-*jet equivalent* at 0, with $r \leq t$, in case the derivatives of $f - g$ of order s vanish for all $s < r+1$. The equivalence class $J_r(f)$ is called the r-*jet* of f at 0.

If $r < \infty$ then there is a *unique polynomial function* in the equivalence class $J_r(f)$, which we denote as f_r. It is the part up to and including degree r of the Taylor series of f at 0. From the definition follows

$J_r(f) = J_r(f_r)$. We will sometimes say "the r-jet f_r" instead of "the r-jet $J_r(f_r)$." Of course it may happen that the functions f_r and f_{r-1} are equal functions, namely when the homogeneous r-th degree part of f_r vanishes. Two C^t-functions f and g are called C^s-*equivalent* or C^s-*diffeomorphic* (or in case $s = 0$ also *homeomorphic*) in case a C^s-diffeomorphism $h: U \to h(U)$ of some neighborhood U of 0 in V onto another h(U) exists, for which

$$f(x) = g(h(x)), \quad x \in U .$$

Following R. Thom we call the r-jet $J_r(f)$ C^s-*sufficient for* C^t-*functions,* if every C^t-function g with the given r-jet:

$$J_r(g) = J_r(f)$$

is C^s-diffeomorphic with f. If $r < \infty$ we have in that case: $J_r(g) = J_r(f) = J_r(f_r)$, and g and f are C^s-diffeomorphic with the polynomial function f_r. We will be mainly interested in such cases. If moreover $t = r$ we say that the r-jet is C^s-*sufficient.*

The differential $df(x) : V_x \to R$ of the C^1-function f at x is a co-vector. It is a linear function on the tangent space V_x at x. Its norm is equal to the length of the gradient vector ${}^t df(x)$ and it will be denoted as

$$|df(x)| = \sqrt{<{}^t df(x), {}^t df(x)>} = \sqrt{<df(x), df(x)>} .$$

Here ^{t} is the canonical isomorphism between an euclidean or Hilbert-space and its dual.

Property Q(r). The c^r-function f has *property* Q(r) $(r < \infty)$, in case $c > 0$ and $\delta > 0$ exist, such that

$$|df(x)| > c|x|^{r-1} \quad \text{for} \quad |x| < \delta . \tag{1}$$

This is in fact a property of the equivalence class $J_r(f)$, in view of

LEMMA 1. *If the* C^r-function f *has property* Q(r), *then so has every* C^r-*function* $g \in J_r(f)$. *In particular* f *has it if and only if* f_r *has it.*

Proof. By the assumptions on g, $df - dg$ is C^{r-1} in x, and its r–1-jet at 0 is a vector valued polynomial function of degree $r-1$, namely the function zero. Hence[1]

$$(df - dg)(x) = 0(|x|^{r-1}) . \qquad (2)$$

Then for any positive $\varepsilon < \frac{1}{2}$ there exists $0 < \delta_1 \leq \delta$, such that

$$|df(x) - dg(x)| < \varepsilon c |x|^{r-1} \quad \text{for} \quad |x| < \delta_1 .$$

With (1) we obtain

$$|dg(x)| > (1-\varepsilon) c\, |x|^{r-1} > \tfrac{1}{2} c\, |x|^{r-1} \quad \text{for} \quad |x| > \delta_1$$

and g has property Q(r).

The *lowest degree* of a function f is the number r_0 for which

$$f_{r_0 - 1} = 0, \quad f_{r_0} \neq 0 .$$

Here 0 means the constant polynomial function 0.

Because $f(0) = 0$, $r_0 \geq 1$. The case $r_0 = 1$ happens if and only if f is not critical at 0, and then by the implicit function theorem the linear function f_1 is C^t-sufficient for C^t-functions. From now on we assume f to be critical at 0:

$$r_0 \geq 2 . \qquad (3)$$

---

1 We will use the *o-symbol* and the *0-symbol of Landau*. For any functions ψ and γ the meaning of the expression

$$\psi(x) = o(\gamma(x))$$

will be

$$\lim_{x=o} \frac{|\psi(x)|}{|\gamma(x)|} = 0 ,$$

where the last expression has to make sense. For example, taking $\gamma(x) \equiv 1$, we see that $\psi(x) = o(1)$ means

$$\lim_{x=o} |\psi(x)| = 0 .$$

On the other hand $\psi(x) = 0(\gamma(x))$ means that constants $c > 0$ and $\delta > 0$ exist, such that

$$|\psi(x) < c|\gamma(x)| \quad |x| < \delta .$$

Occasionally we will allow ourselves to substitute $o(\gamma(x))$ or $0(\gamma(x))$ for $\psi(x)$ in these respective cases. We will also use some obvious properties of this symbolism.

We can now formulate our main theorems.

THEOREM 1. *Let* $r_o \geq 2$ *be the lowest degree of the* C^r*-function* $f: V \to R$, $f(0) = 0$. *If* f *has property* $Q(r)$, *then* f_r *is* C^o*-sufficient for* C^r*-functions* $(r < \infty)$.

THEOREM 2. *If the* C^{r+p}*-function* $f: V \to R$, $f(0) = 0$, *has property* $Q(r)$ *and* $p \geq r - r_o$, *then* $J_{r+p}(f)$ *is* C^1*-sufficient for* C^{r+p}*-functions.*

In particular for $p < \infty$ any C^{r+p}-function in $J_{r+p}(f)$ is C^1 diffeomorphic with the polynomial function f_{r+p}.

COROLLARY. *The special case* $r = r_o$. *Let* f *be* C^{r+p}. *If* f_r *is a homogeneous polynomial function of degree* r *and it has property* $Q(r)$, *then* f_r *is* C^1*-sufficient.*

It is easily seen that a function f with critical point 0 can be defined to be non-degenerate, and this has the usual meaning, also for Hilbert-space, if and only if it has property $Q(2)$. Then $r = r_o = 2$. We read from Theorem 2 that any C^2-function with non-degenerate isolated critical point at 0 is C^1-diffeomorphic with the polynomial function f_2.

In (1) we proved more general

THEOREM 3 "Morse Lemma." *Every homogeneous non-degenerate quadratic function* $f_2 : V \to R$ *is* C^{p+1}*-sufficient for* C^{p+2}*-functions for* $0 \leq p \leq \infty$.[2]

---

[2] The classical Morse lemma concerned finite dimensional space $V = R^n$, the loss of differentiability was two and it applied only to functions which were at least C^3. Palais [8] gave a proof to cover the Hilbert-space case too. See Appendix §5. In [1] we obtained Theorem 3 with special explicit methods. Ostrowski [7] gave a simple proof of Theorem 3 for finite dimensional V.

§2. *Discussion of the property* Q(r)

If $f: V \to R$ has at $0 \in V$ the property Q(r) with $2 \leq r < \infty$, then f has an isolated critical point at 0. The converse is not true as can be seen from the C^∞-function $f: R^2 \to R$, defined by

$$f(x_1, x_2) = \begin{cases} x_1^2 + \exp - x_2^{-2} & x_2 \neq 0 \\ x_1^2 & x_2 = 0 \end{cases}$$

which has a minimum at 0. There is no $r < \infty$ such that f_r is C^0-sufficient. Indeed for any $r < \infty$ the function

$$\begin{cases} f(x_1, x_2) - 2 \exp - x_2^{-2} & x_2 \neq 0 \\ x_1^2 & x_2 = 0 \end{cases}$$

has the same r-jet at 0 as f, but it has no minimum at $0 \in R^2$. In this example f_r has at 0 no isolated critical point for any $r < \infty$. If f_r has an isolated critical point, then again the property Q(r) need not hold. For example take the function g:

$$g(x_1, x_2) = (x_2 - x_1^2)^2 + x_2^4 .$$

It equals g_r for $r = 4$ and has the isolated critical point 0. For small $t > 0$ we find at the point $x = (x_1, x_2) = (t, t^2)$:

$$|x| = t\sqrt{1+t^2}, \quad |dg(x)| = 4t^6 .$$

Hence g does not have property Q(4). It does obey property Q(7) however. If f is a real analytic function on a finite dimensional space $V = R^n$, then $|df(x)|^2$ is also analytic. Suppose it vanishes in some neighborhood of $0 \in R^n$ only at 0. Then by the inequality of Lojaciewicz [2] there exists an integer r and $\delta > 0$, $c > 0$ such that

$$|df(x)|^2 > c|x|^{2r} \quad \text{for} \quad |x| < \delta .$$

Consequently:

THEOREM 4A. *If* $V = R^n$ *has finite dimension and* $f: V \to R$ *is real analytic with an isolated critical point at* 0, *then* f *has property* Q(r) *for some* $2 \leq r < \infty$.

Next we prove

THEOREM 4B. *If* $f: R^n \to R$ *is such, that* f_r *is homogeneous of degree* r *with isolated critical point* 0, *then* f *has property* Q(r).

Proof. $|df_r(x)|^2$ is a homogeneous polynomial function of degree $2r-2$. Hence for $t > 0$

$$|df_r(tx)|^2 = t^{2r-2}|df_r(x)|^2 .$$

The function increases along each half ray that starts at 0. By continuity, a number $\delta > 0$ exists, such that

$$|df_r(x)| \geq 1 = c|x|^{r-1} \quad \text{for} \quad |x| = \delta,\ c = \delta^{1-r} .$$

As $|df_r(x)|$ is homogeneous of degree r–1, then Q(r) follows:

$$|df_r(x)| \geq c|x|^{r-1} \quad \text{for} \quad x \in V .$$

For Hilbert-space H there is the following:

Counterexample 4C. *The function* $f \in H \to R$ *defined by*

$$f(x) = \Sigma_1^\infty x_j^4$$

on real Hilbert-space $H = \{x = (x_1, x_2, \ldots) : \Sigma_1^\infty x_j^2 < \infty\}$ *has an isolated critical point at* $0 \in H$. *It has property* Q(r) *for no* r *whatsoever.*

Proof. Suppose for some r and $\delta > 0$, $c > 0$

$$|df(x)| > c|x|^{r-1} \quad \text{for} \quad |x| \leq \delta . \tag{4}$$

The differential of f at x is the linear function in $y \in H$:

$$df(x) : y \to \Sigma_1^\infty 4x_j^3\, y_j .$$

It has the norm

$$|df(x)| = \sqrt{\Sigma_1^\infty (4x_j^3)^2} \ .$$

Take

$$x_i = \frac{\delta}{\sqrt{m}} \text{ for } i \leq m, \quad x_i = 0 \text{ for } i > m \ .$$

Then $|x| = \delta$, $|df(x)| = \frac{4\delta^3}{m}$. And this is smaller than $c\,|x|^{r-1} = c\delta^{r-1}$ for m sufficiently large, contradicting assumption (4).

THEOREM 4D. *Let* $f: V \to R$ *be a* C^r*-function and* r *odd. Suppose (condition* $P(r)$*) there exists a* C^1*-map* $\phi : x \to \phi_x \in \lim(V', R) = V''$, *such that for some neighborhood* U *of* 0 *in* V:

$$|x|^{r-1} = \phi_x(df(x)) \quad \text{for} \quad x \in U \ . \tag{4}$$

Then f *has property* $Q(r)$.

Remark. For finite dimensional space $V = R^n$ the condition of the theorem holds in case the partial derivatives $(\partial_i f_r)(x)$ $(i = 1, \ldots, n)$ generate an ideal in the ring of formal power series that contains all homogeneous polynomials of homogeneous degree $r-1$, *hence in case it contains the* $r-1$-th *power* μ^{r-1} *of the maximal ideal* μ. Such conditions are used in the profound work of Tougeron [9] and Mather [3] which is concerned with C^∞-equivalence of functions and ideals of functions near 0.

Proof. If the property $Q(r)$ does not hold, then a sequence of points $x_{(j)}$ $(j = 1, 2, \ldots)$ converging to $0 \in V$ exists, such that

$$|(df)(x_{(j)})| < \frac{1}{j}|x_{(j)}|^{r-1} \ .$$

As ϕ is C^1 there is a constant c such that

$$|\phi_x(y)| < c \cdot |y| \ \text{ for small } |y| \ .$$

Hence

$$|\phi_x(df)(x_{(j)})| < \frac{c}{j}|x_{(j)}|^{r-1} \ \text{ for large } j \ ,$$

and by (4) we have the contradiction

$$|x_{(j)}|^{r-1} < \frac{c}{j}|x_{(j)}|^{r-1} \text{ for large } j .$$

§3. *Sufficiency and non-sufficiency*

In this section we give examples in order to illustrate the main Theorem 2, as well as counterexamples.

THEOREM 5A. *Let* $f_r : R^n \to R$ *be a homogeneous polynomial of degree* r, *which is invariant under at most a finite number of linear automorphisms of* R^n. *Then* f_r *is not* C^2*-sufficient, not even for polynomial functions, in the cases:*

$$r = 3,\ n \geq 7;\ r = 4,\ n \geq 3;\ r \geq 5,\ n \geq 2 . \tag{5}$$

Proof. The number of parameters needed to get every $r+1$-jet that has a given r-jet, is equal to the dimension of the linear space of homogeneous forms of degree $r+1$ in $x_1, \ldots, x_n$:

$$\binom{n + (r+1) - 1}{r+1} = \binom{n+r}{1+r} .$$

Substituting $x_i' = \Sigma_j\, \alpha_{ij} x_j + \Sigma_{j,k}\, \alpha_{ijk} x_j x_k$ instead of x_i in the polynomial f_r, we get the old r-jet f_r back for a finite number of choices of the matrix $\{\alpha_{ij}\}$. The dimension of the available parameter space is therefore equal to the number of coefficients α_{ijk}, $j \leq k$. This number is $\frac{1}{2}n^2(n+1)$. Therefore not all $r+1$-jets can be obtained by a C^2-transformation of coordinates, if

$$\frac{n^2(n+1)}{2} < \binom{n+r}{1+r}$$

This is condition (5).

If we allow f_r to be invariant under a subgroup of $GL(n, R)$ of dimension greater than 0, we get a slightly weaker condition than (5).

Example. $x_1^5 + x_2^5 + x_1^3 x_2^3$ and $y_1^5 + y_2^5$ are C^1-diffeomorphic by Theorems 2 and 4B. They are not C^2-diffeomorphic as one sees by substituting power series in x_1, x_2 modulo terms of degree greater than two for y_1 and y_2. The term $x_1^3 x_2^3$ cannot be obtained.

Example. $x_1^3 + x_2^3 + \dots + x_7^3$ is C^1-sufficient for C^3-functions. It is not C^2-sufficient for polynomials.

Remark. The function in two variables $x_1^5 - x_2^5$ is C^1-diffeomorphic with $x_1^5 - x_2^5 + x_1^3 x_2^6$. However, Tougeron and Mather can assert much better namely C^∞-diffeomorphism.

THEOREM 5B. *The functions* $(x_1^2 + x_2^2)^2 + x_1^{5+p}$ *and* $(x_1^2 + x_2^2)^2$, $p \geq 0$, *are* C^1-*diffeomorphic and not* C^{2+p}-*diffeomorphic.*

The interest in this theorem lies in the fact that it holds for arbitrarily large p, for example $p = 995$.

Proof. The first statement follows from Theorem 2. Next suppose the given functions are C^{2+p}-diffeomorphic. Then there exist C^{2+p}-functions $y_i = h_i(x_1, x_2)$, $i = 1, 2$, where Taylor series start with terms of degree one, such that

$$(x_1^2 + x_2^2)^2 + x_1^{5+p} = (y_1^2 + y_2^2)^2 = \left[h_1^2(x_1, x_2) + h_2^2(x_1, x_2)\right]^2 \equiv \phi^2(x_1, x_2)\,.$$

The lowest and the highest relevant degree of Taylor series is

for	lowest	highest relevant
h_i	1	$2+p$
h_i^2	2	$3+p$
ϕ^2	4	$5+p$
$\frac{\partial \phi^2}{\partial x_2}$	3	$4+p$
$\left(\frac{\partial \phi^2}{\partial x_2}\right)^2$		$7+p$

Now we have
$$\left(\frac{\partial \phi^2}{\partial x_2}\right)^2 = \left(2\phi \frac{\partial \phi}{\partial x_2}\right)^2 = \phi^2 \cdot 4\left(\frac{\partial \phi}{\partial x_2}\right)^2 . \tag{6}$$

The left side is $\left[4x_2\left(x_1^2+x_2^2\right)\right]^2$.

The right side is $\left[\left(x_1^2+x_2^2\right)^2+x_1^{5+p}\right]\cdot 4\left(\frac{\partial\phi}{\partial x_2}\right)^2$.

Equality of the beginning of the Taylor series yields

$$4\left(\frac{\partial\phi}{\partial x_2}\right)^2 = 16x_2^2 \quad \text{modulo terms of higher order.}$$

Then we obtain in polynomials modulo terms of order $8+p$ from (6):

$$\left[4x_2\left(x_1^2+x_2^2\right)^2\right]^2 = \left[\left(x_1^2+x_2^2\right)^2+x_1^{5+p}\right]\left[16x_2^2+A(x_1,x_2)\right].$$

$A(x_1, x_2)$ is a polynomial of degree $3+p$ and lowest degree ≥ 3. But then inductively this lowest degree is seen to be $> 3+p$; that is $A(x_1, x_2) = 0$. Then the terms of degree $7+p$ give a contradiction.

Example 5C. The function defined on real Hilbert space $f: H \to \mathbf{R}$, by

$$f(x) = f(x_1, x_2, \ldots) = \Sigma_1^\infty x_j^4 - \left(\Sigma_1^\infty x_j^2\right)^3$$

has the properties:

a. For every finite dimensional subspace $W \subset H$, the restriction $f|W$ has property $Q(4)$. It has an isolated critical point $0 \in W$ with minimal value 0.

b. In every neighborhood of $0 \in H$ the function has negative values. Compare counterexample 4C.

§4. *Proof of the main Theorems 1 and 2*

It is sufficient to prove

THEOREM A. *Let the* C^{r+p}*-function* $f:(V, 0) \to (\mathbf{R}, 0)$ *with smallest degree* $r_o \geq 2$ *have property* $Q(r)$ $(p \geq 0)$. *Hence* $c > 0$ *and* $\delta > 0$ *exist such that*

$$|df_{r+p}(x)| > c|x|^{r-1} \quad \text{for} \quad x < 2\delta\,. \tag{7}$$

a. Then there is a homeomorphism $h: U \to h(U)$, U and $h(U)$ neighborhoods of 0 in V, such that

$$f_{r+p}(h(x)) = f(x) \quad \text{for} \quad x < U\,. \tag{8}$$

In case $p = \infty$, *then* f_{r+p} *will stand for some arbitrary* C^∞*-function having all derivatives at* 0 *in common with* f.

b. The restriction of h to $U - \{0\}$ is a C^{r+p}-diffeomorphism.

c. If $p \geq r - r_o$, then h is a C^1-diffeomorphism.

Indeed for $p = 0$ the first consequence a. yields Theorem 1. The parts b. and c. then yield Theorem 2.

The different steps in the proof will be emphasized by the formulation of lemmas which will eventually together constitute the proof of Theorem A. We begin with a preliminary

LEMMA 2. *If the* C^r*-functions* f *and* g *have the same* r*-jet and obey condition* $Q(r)$, *then for any positive* $\varepsilon < \pi/2$, *there is* $\delta > 0$ *such that*

$$\text{angle } ({}^t df(x),\ {}^t dg(x)) < \varepsilon \quad \text{for} \quad |x| < \delta\ .$$

In particular the orthogonal trajectories of f are transversal to the level hyper-surfaces of g for $0 < |x| < \delta$.

Proof. We consider the difference of the unit vectors in the directions of the gradients of f and g at x, and observe:

$$\left| \frac{df(x)}{|df(x)|} - \frac{dg(x)}{|dg(x)|} \right| \leq \frac{|df(x) - dg(x)|}{|df(x)|} + |dg(x)| \left| \frac{1}{|df(x)|} - \frac{1}{|dg(x)|} \right|$$

$$\leq 2\, \frac{|df(x) - dg(x)|}{|df(x)|}\ .$$

By (1) $(Q(r))$ and (2) this is smaller than $\frac{1}{2}\varepsilon$ for sufficiently small x, say $|x| < \delta$. Then the lemma follows.

From now on f is a function obeying the assumptions of Theorem A.

Definition. Any connected orthogonal trajectory of the level hyper-surfaces of the function f_{r+p}, oriented by increasing f_{r+p}-values, will be called a *path*. By Lemma 2 we may assume δ so small, that the function f increases along any path in the ball $B(2\delta) = \{x|\ |x| < 2\delta\}$. Such a path meets any level hyper-surface of the function f at most once. We now define $h(x)$ for $|x| < \delta$ as the unique (if existing) intersection point of the path through x and the level $f_{r+p}^{-1}(f(x)) \cap B(2\delta)$. Hence, if we succeed, we have the required condition:

$$f_{r+p}(h(x)) = f(x) . \tag{8}$$

For $x = 0$ we define $h(0) = 0$.

Our first aim will be

LEMMA 3. *For sufficiently small* $\delta > 0$, h *is a well defined homeomorphism.*

Proof. We take a point $x_o \neq 0 \in V$. If $f_{r+p}(x_o) = f(x_o)$ then $h(x_o) = x_o$, and by continuity $h(x)$ is well defined in some neighborhood of x_o. So we may assume $f_{r+p}(x_o) \neq f(x_o)$. We even assume (the other case goes analogously)

$$f_{r+p}(x_o) < f(x_o) . \tag{9}$$

Let

$$|x_o| = b < \delta, \quad 0 \leq s \leq \frac{b}{r-1} = \frac{|x_o|}{r-1} . \tag{10}$$

The path through x_o with arclength s increasing with the value of f_{r+p}, is the solution $x(s)$ of the differential equation:

$$\frac{dx}{ds} = \frac{{}^t df_{r+p}(x)}{|{}^t df_{r+p}(x)|} \qquad x(0) = x_o . \tag{11}$$

It is well defined for $s < \frac{b}{r-1} < \delta$ because the path does not get long enough in order to disappear from $B(2\delta)$ or hit $0 \in V$. With (11) and (7) we find

$$\left|\frac{df_{r+p}(x(s))}{ds}\right| = |df_{r+p}(x(s))| > c\,|x(s)|^{r-1} .$$

If we replace for all s the point $x(s)$ by the point $\left(1 - \frac{s}{b}\right)x_o$, then the new points form a straight arc that starts at x_o and is directed towards $0 \in V$. The value of the function $|x|^{r-1}$ along this arc, which also has s as arclength, diminishes with respect to s, at least as fast as along $x(s)$ as we see from a figure of the two arcs.

Therefore

$$f_{r+p}(x(s)) - f_{r+p}(x_o) = \int_o^s \frac{df_{r+p}(x(s))}{ds} ds \geq c \int_o^s |x(s)|^{r-1} ds \geq$$

$$c \int_o^s \left\{\left(1 - \frac{s}{b}\right) b\right\}^{r-1} ds = c \int_o^s (b-s)^{r-1} ds$$

$$= \frac{c}{r}(b^r - (b-s)^r) \geq \frac{c}{2} b^{r-1} s \quad \text{for} \quad s \leq \frac{b}{r-1} .$$

Consequently:

$$f_{r+p}(x(s)) - f_{r+p}(x_o) \geq \frac{c}{2} |x_o|^{r-1} s \quad \text{for} \quad s \leq \frac{|x_o|}{r-1} . \tag{12}$$

We put $p' = p$ for $p < \infty$ and p' any integer ≥ 0 for $p = \infty$. Because

$$f(x_o) - f_{r+p}(x_o) = 0(|x_o|^{r+p'}) ,$$

there exists for any positive $\varepsilon < 1$ a positive $\delta(\varepsilon) \leq \delta$ so small that

$$0 < |f(x_o) - f_{r+p}(x_o)| < \frac{\varepsilon c}{4(r-1)} |x_o|^{r+p'} \quad \text{for} \quad |x_o| < \delta(\varepsilon) . \tag{13}$$

This extra condition $|x_o| < \delta(\varepsilon)$ depends on p'. We could have stated it right at the beginning of §4!

For
$$s = s_1 = \frac{\varepsilon}{2(r-1)} |x_o|^{1+p'} \leq \frac{|x_o|}{r-1} , \tag{14}$$

the right hand sides of (12) and (13) are equal, and we obtain a long sequence of inequalities:

$$0 < f(x_o) - f_{r+p}(x_o) < \frac{c}{2} |x_o|^{r-1} s_1 \leq f_{r+p}(x(s_1)) - f_{r+p}(x_o) .$$

Hence

$$f_{r+p}(x_o) < f(x_o) < f_{r+p}(x(s_1)) .$$

The mounting path through x_o therefore has at "time" s_1 already passed beyond a "time" $s < s_1$ at which $f_{r+p}(x(s)) = f(x_o)$. Then that point $x(s)$ exists!

Taking $\varepsilon = 1$, we see that $h(x_o)$ is now well defined for $|x_o| < \delta(1)$. The length of the straight segment between x_o and $h(x_o)$ is smaller or equal to the arclength. Hence

$$|h(x_o) - x_o| \leq s \leq s_1 = \frac{\varepsilon}{2(r-1)} |x_o|^{1+p'} \leq \varepsilon |x_o|^{1+p'} \quad \text{for} \quad |x_o| < \delta(\varepsilon) .$$

Consequently

$$h(x) - x = o(|x|^{1+p'}), \quad p' \begin{cases} = p & \text{for} \quad p \neq \infty \\ < \infty & \text{for} \quad p = \infty \end{cases} . \tag{15}$$

Recalling that $h(0) = 0$ we conclude that h *is continuous* at $x = 0$. Denoting the *identity automorphism* of V by e, we obtain from (15)

$$h(x) - h(0) - ex = o(|x|) .$$

Consequently the *derivative of* h *at* 0 *is* e, by the definition of derivative.

We have used our paths to carry a point with a particular f-level value to the point with that same f_{r+p}-level value. We can use the paths also for the opposite purpose and obtain for δ small enough the inverse h^{-1} of h. This implies Lemma 3 and part a. of Theorem A. Part b. is:

LEMMA 4. h *is* C^{r+p} *for* $0 < |x| < \delta(1)$.

Proof. In some neighborhood of x_o $(0 < |x_o| < \delta(1))$ we use the function f_{r+p} as one coordinate and we coordinatize the points further by their paths which again can be represented by points in the f_{r+p} level hypersurface through x_o. Near $h(x_o)$ we use the function f and the paths. f is C^{r+p}. h is now represented by the identity map in these two sets of coordinates. Hence as f is C^{r+p}, also h is C^{r+p}.

LEMMA 5. *If* $p \geq r - r_o$ *then*

$$dh(x_o) - e = o(1) ,$$

hence h *is* C^1 *also at* $x_o = 0$, *and Theorem A is proved.*

Let $p \geq r - r_o$. (16)

We study h in some very small neighborhood of a point $x_o \neq 0$, near to 0.

Let

$$\mathbf{n} = {}^t df_{r+p}(x_o)/|df_{r+p}(x_o)|$$

be a unit normal vector orthogonal to the f_{r+p}-level at x_o. We introduce new orthonormal "coordinates" (u, v) as follows.

$$v(x) = df_{r+p}(x_o) \cdot (x - x_o)/|df_{r+p}(x_o)|$$

is the distance of x (up to sign) to the tangent hyperplane at x_o with respect to its f_{r+p}-level. Hyperplanes parallel to this hyperplane will be called horizontal.

$$u(x) = x - x_o - v(y) \cdot \mathbf{n}$$

is a vector in the vector subspace orthogonal to $\mathbf{n}$. The new coordinates of x_o are (0, 0).

For x near to x_o the path of x meets the coordinate plane $v = 0$ in exactly one point to be denoted by p(x). The function p(x) is C^∞.

Next we define a map κ, for points near to x_o in terms of the new coordinates, by

$$\begin{cases} u(\kappa(x)) = u(p(x)) \\ v(\kappa(x)) = \dfrac{f_{r+p}(x) - f_{r+p}(x_o)}{|df_{r+p}(x_o)|} . \end{cases}$$

The C^∞-map κ carries paths onto vertical lines. It has the properties:

$$\kappa(x_o) = x_o, \qquad d\kappa(x_o) = e .$$

Then κ is a C^∞-diffeomorphism near x_o.

Now the map h can be factorized near x_o as follows

$$x \to \kappa(x) \to \kappa(h(x)) \to h(x) . \tag{17}$$

It remains to prove that the first derivatives of these three maps tend to e for $x_o \to 0$ and

$$|x - x_o| < 2|h(x_o) - x_o| .$$

In order to study the first derivative of κ near x_o, we use a microscope ϕ:

$$\phi: x \to y = \frac{x - x_o}{\alpha}$$

$$\phi^{-1}: y \to x = x_o + \alpha y$$

with enlargement α^{-1}, where α is the constant

$$\alpha = \alpha(x_o) = 2|h(x_o) - x_o| + |x_o|^{2+p} = o(|x_o|^{1+p}). \tag{19}$$

Compare (16). The term $|x_o|^{2+p}$ is added because $h(x_o) - x_o$ may happen to be zero.

Under the microscope the distance between x_o and $h(x_o)$ is seen as

$$\frac{|h(x_o) - x_o|}{\alpha(x_o)} \leq ½ .$$

We study the function

$$y \to g_{r+p}(y) = \frac{f_{r+p}(x_o + \alpha y)}{\alpha|df_{r+p}(x_o)|}$$

which at $y = 0$ has a gradient of length one. The orthogonal trajectories of the levels of g_{r+p} are the microscope-images of the paths. We will see that these trajectories are close to vertical lines for small x_o.

We want to estimate

$$d^2 g_{r+p}(y) = \frac{\alpha^2\, d^2 f_{r+p}(x)}{\alpha|df_{r+p}(x_o)|}$$

With (19), the condition $Q(r)$, the fact that the Taylor series of $d^2 f_{r+p}(x)$ has lowest degree $r_o - 2$ in x, and x is near to x_o (say $|x| < 2|x_o|$) one obtains:

$$d^2 g_{r+p}(y) = o(|x_o|^{1+p}) \cdot 0(|x_o|^{r_o - 2}) \cdot |x_o|^{-(r-1)}$$

$$= o(|x_o|^{p-(r-r_o)}) = o(1) \qquad \text{by (16).}$$

Observe that the condition (16) $p \geq r - r_o$ is essentially needed here. In the domain $|y| < 1$ the function $y \to g_{r+p}(y)$ is arbitrarily close to a linear function with gradient of length one, for small $|x_o|$.

By integration one finds for $|y| < 1$ that the mapping $\phi\kappa\phi^{-1}$, its first derivative included, is arbitrarily close to the identity map. But ϕ is just a similarity transformation. Therefore the first derivative of κ is also arbitrarily close to e for $|x-x_o| < a = a(x_o)$ and for small $|x_o|$.

Consequently, as $|h(x_o) - x_o| < \frac{1}{2}a$, we may conclude that the first and the last map in (17) have both first derivatives arbitrarily close to e for $x = x_o$ and small $|x_o|$.

The second map in (17) can be expressed conveniently in terms of the coordinates (u,v). It carries the point $\kappa(x)$ with coordinates $(u(\kappa(x)), v(\kappa(x)))$ into the point $\kappa(h(x))$ with coordinates

$$\left(u(\kappa(x)),\ v(\kappa(x)) + \frac{f_{r+p}(h(x)) - f_{r+p}(x)}{|df_{r+p}(x_o)|}\right).$$

The differential of the difference in the (only relevant) second (one-dimensional) coordinate is (with respect to x)

$$\frac{d(f(x) - f_{r+p}(x))}{|df_{r+p}(x_o)|} = \frac{o(|x_o|^{r+p-1})}{|x_o|^{r-1}} = o(|x_o|^p) = o(1)\ .$$

Hence also the derivative of the second map in (17) is arbitrarily close to e for small $|x_o|$.

Then the derivative at x_o of the map h is close to e for small $|x_o|$ and Lemma 5 is proved.

§5, *Appendix. Palais' proof of the Morse Lemma*

Let $f : V \to R$ be a C^{3+p}-function with non-degenerate critical point $0 \in V$; $f(0) = 0$; $0 \le p \le \infty$. We prove the existence of a C^{1+p}-diffeomorphism $h : U \to h(U)$ such that

$$f(x) = f_2(h(x)) \quad \text{for} \quad x \in U\ .$$

Proof. As $f(0) = 0$ we have (see Milnor [4])

$$f(x) = \int_o^1 \frac{d}{dt} f(tx)\,dt = \int_o^1 (df)(tx)\,dt \cdot x = g(x)\cdot x\ .$$

Here $g(x) = \int_0^1 (df)(tx)dt \in \text{Lin}(V, \mathbb{R}) = V'$. V' is the dual space of V. $g: V \to V'$ is a C^{2+p}-map with $g(0) = (df)(0) = 0$. So we lost one class of differentiability. This happens again in the next step:

$$g(x) = \int_0^1 (dg)(tx)dt \cdot x = k(x) \cdot x \ .$$

Here $k(x) \in \text{Lin}(V, V')$, and $k: V \to \text{Lin}(V, V')$ is a C^{1+p}-map. Substitution yields

$$f(x) = (k(x) \cdot x) \cdot x \ .$$

We replace the bilinear function

$$k(x) : (y, z) \to (k(x) \cdot y) \cdot z$$

by the symmetric function which assigns to (y, z) the value

$$\tfrac{1}{2}[(k(x) \cdot y) \cdot z + (k(x) \cdot z) \cdot y] = {}^t y H_x z \ .$$

The last equality is by definition of $H_x \in \text{Lin}(V, V)$. Observe

$$f_2(x) = {}^t x H_o x$$

$$f(x) = {}^t x H_x x$$

and the matrices (operators) H_o and H_x are symmetric (self-dual):

$${}^t H_o = H_o, \quad {}^t H_x = H_x \ .$$

Let B_x be defined by $H_x = H_o B_x$.

Then

$${}^t H_x = {}^t B_x \, {}^t H_o = {}^t B_x H_o = H_o B_x \ .$$

For small $|x|$, say $|x| < \delta$, the operator B_x is close to 1, say

$$|B_x - 1| < \varepsilon < 1 \ .$$

Hence $C_x = \sqrt{B_x} = \sqrt{1 + (B_x - 1)}$ can be defined by its uniformly convergent power series in $B_x - 1$. Then $C_o = 1$ and C_x depends of class C^{1+p} on x.

Observe also that

$$^tC_x H_o = H_o C_x .$$

Then

$$f(x) = {}^tx\, H_x x = {}^tx\, H_o\, C_x^2 x = {}^tx\, {}^tC_x\, H_o\, C_x x$$

$$= {}^t(C_x x)\, H_o\, (C_x x) = f_2(h(x)) ,$$

where $h(x) = C_x x$ is the required C^{1+p}-diffeomorphism defined for $x \in U$. U is a sufficiently small neighborhood of $0 \in V$.

Remark 1. For $p = -1$, C_x can still be defined in the same manner, but one cannot decide that h has an inverse near $x = 0$.

Remark 2. There exists a linear "change of coordinates" $\sigma: V \to V$, and a linear subspace $V_1 \subset V$ with orthogonal complement $V_2 \subset V$, such that for $u(y) \in V_1$ and $v(y) \in V_2$, $y = u(y) + v(y)$, the function has the standard form $|u|^2 - |v|^2$ for quadratic functions on V:

$$f(x) = f_2(h(x)) = |u(\sigma hx)|^2 - |v(\sigma hx)|^2 ,$$

(See Palais [8] for the Hilbert space.)

REFERENCES

[1] N. H. KUIPER – C^r-*functions near non-degenerate critical points*, Mimeographed, Warwick University, Coventry 1966.

[2] B. MALGRANGE – *Seminaire Cartan* 1962/63, no. 11, 12, 13, 22.

[3] J. MATHER – Thesis, Princeton, N. J. Work to appear in *Ann. of Math.*

[4] J. MILNOR – Morse theory, *Ann. Math. Studies* 51, p. 6.

[5] M. MORSE – Relations between the critical points of a real function of n variables, *Trans. Amer. Math. Soc.* 27 (1925), 345-396.

[6] M. MORSE – The reduction of a function near a non-degenerate critical point, *Proc. Nat. Acad. Sci.* U.S.A. 54 (1965), 1795-1763.

[7] A. M. OSTROWSKI,– On the Morse-Kuiper theorem, to appear in "*On Functional Equations.*"

[8] R. PALAIS – Morse theory on Hilbert manifolds, *Topology* 2 (1963), 299-340.

[9] J. C. TOUGERON – Thèse à Rennes. Compare *C. R. Acad. Sci.* Paris 262 (A et B) (1966), 563-565.

UNIVERSITY OF AMSTERDAM

FIXED POINT INDEX AND LEFSCHETZ NUMBER

by

JEAN LERAY

Let E be a topological space, 0 an open subset of E, $\bar{0}$ its closure and ξ a continuous mapping: $\bar{0} \to E$; its fixed points are the points $x \in \bar{0}$ such that $x = \xi(x)$.

Twenty years ago I gave in [6] a new definition of their index $i_\xi(0)$; it has the following features: it does not assume E linear; it is independent of the notion of topological degree [1]; it has close relations with Lefschetz number (see No. 8, below). Few years later I improved and developed in [7] some of the tools used by the preceding definition: fine covertures (i.e., simple and full gratings), spectral sequence and filtrated cohomology. This second paper made possible an important simplification of that new definition of the fixed-point index. Recently, this simplification was explicitly exposed by D. G. Bourgin in his excellent *Modern Algebraic Topology* [1]; he made some change of notation and terminology (his gratings are my complexes; his simple gratings my covertures; his full gratings my fine complexes).

The last chapter of [6] (Ch. VII) gives an extension of the notion of fixed-point index which deserves a similar simplification; the present report paves the way for it by describing as shortly as possible the main features of the fixed-point theory itself: it sums up [6], [7], [8], [3]; it has relations with the reports by A. Granas [4], R. Knill [5] and also with F. E. Browder's paper [2].

---

1 Whereas in [10] J. Schauder and myself assumed E linear and defined that index as the topological degree of mapping: $x \to x - \xi(x)$.

My purpose being to make accessible the part of [6] which is not presented by D. G. Bourgin [1], I must keep my terminology, but I shall quote his own.

§1. COHOMOLOGY OF COMPACT SPACES

The following definition of the cohomology of a compact[2] space is an extension of de Rham's definition [11] of the cohomology of differentiable manifolds as the cohomology of the algebra of their exterior differential forms (E. Cartan suggested that definition in 1928 after he succeeded in understanding a sentence written by H. Poincaré in 1899).

1. COMPLEX (Grating in [1]). Let E be a compact space.

Definition. A complex (or grating) C of E is a graded differential ring with a unity U and supports S(...), contained in E.

Let us explain the meaning of these terms:

"Graded" means that

$$C = \underset{p}{\otimes} C^p \quad (p: \text{ integer } \geq 0)$$

is the direct sum of additive groups C^p without torsion[3]; the elements $X^{p,\alpha}$ of C^p are said to be homogeneous of degree p; the degree of a product is the sum of the degrees of its factors:

$$X^{p,\alpha} . X^{q,\beta} \in C^{p+q} \quad (\text{obviously } U \in C^0) .$$

"Differential" means that a linear mapping is given

$$d: \; C^p \to C^{p+1}, \;\; \forall p$$

such that

$$d^2 = 0, \;\; d(X^{p,\alpha} . X^{q,\beta}) = (dX^{p,\alpha}) \cdot X^{q,\beta} + (-1)^p X^{p,\alpha} \cdot dX^{q,\beta}$$

(obviously $dU = 0$ since $U = U^2$ and $U \in C^0$); the mapping d is called the differential of C.

---

2 It can be adapted to locally compact spaces.

3 i.e.: If $X \in C$ and if $m = \text{integer} \neq 0$, then $mX = 0$ implies $X = 0$.

"With supports" means that to each $X \in C$ is associated a closed subset $S(X)$ of E such that:

$$S(X) = \emptyset \text{ if and only if } X = 0; \quad S(mX) = S(X) \text{ for } m = \text{integer} \neq 0;$$

$$S(\Sigma_p X^p) = \cup_p S(X^p) \text{ if } X = \Sigma_p X^p, \; X^p \in C^p;$$

$$S(X^{p,\alpha} + X^{q,\beta}) \subseteq S(X^{p,\alpha}) \cup S(X^{q,\beta}); \; S(X^{p,\alpha} \cdot X^{q,\beta}) \subseteq S(X^{p,\alpha}) \cap S(X^{q,\beta});$$

$$S(dX) \subseteq S(X).$$

Definition. C is *basic* if the additive group C has a finite base[4] $\{X^{p,\alpha}\}$, such that

$$S(\Sigma_{p,\alpha} c_{p,\alpha} X^{p,\alpha}) = \cup_{p,\alpha} S(X^{p,\alpha}) \quad (c_{p,\alpha}\text{: integers} \neq 0).$$

C is *fine* (full in [1]) when the unity U of C has partitions with arbitrarily small supports; i.e.: an open cover of E being given, there exists a partition of U:

$$U = \sum_\alpha X^{0,\alpha},$$

such that each $S(X^{0,\alpha})$ belongs to some element of that cover.

Of course a fine complex cannot be basic (except if E is a finite set of points). The definition of the cohomology uses fine complexes; instead of them basic complexes can be used, under convenient assumptions, which play a fundamental role in fixed-point theory.

The definition of the cohomology uses also the following operations:

Operations on a complex. Let ξ be a continuous mapping:

$$\xi\colon E' \to E \quad (E' \text{ and } E\text{: compact spaces});$$

---

[4] Each $X \in C$ has a unique expression $X = \Sigma_{p,\alpha} c_{p,\alpha} X^{p,\alpha}$, where $c_{p,\alpha}$ are integers $\gtrless 0$.

denote by C' the differential ring isomorphic to C, with supports $\xi^{-1}S(X)$; its elements with empty supports constitute an ideal I' of C'; the quotient ring C'/I' is obviously a complex of E'; it is denoted by $\xi^*(C)$; $\xi^*(X)$ denotes the image of $X \in C$.

That complex is denoted by $C \cdot E'$ and called the section of C by E' when $E' \subset E$, ξ being the inclusion mapping: $E' \to E$; $\xi^*(X)$ is denoted by $X \cdot E'$; thus

$$S(X \cdot E') = S(X) \cap E', \quad \forall X \in C .$$

Let C and C' be two complexes of E; their tensorial product $C \otimes C'$ is obviously a graded differential ring; we identify $C \otimes C'$ with $C' \otimes C$ by identifying $X^{p,\alpha} \otimes X'^{q,\beta}$ with $(-1)^{pq} X'^{q,\beta} \otimes X^{p,\alpha}$; define as follows supports of its elements:

$$x \in S\left(\sum_\alpha X^{p,\alpha} \otimes X'^{q,\alpha}\right) \text{ if and only if } \sum_\alpha (X^{p,\alpha} \cdot x) \otimes (X'^{q,\alpha} \cdot x) \neq 0$$

in $(C \cdot x) \otimes (C' \cdot x)$; $(p = p(\alpha),\ q = q(\alpha))$; obviously the elements with empty support constitute an ideal I of $C \otimes C'$; the quotient $(C \otimes C')/I$ is a complex of E denoted by $C \circ C'$; $C \circ C'$ is called the intersection of C and C'. The image of $\Sigma_\alpha X^{p,\alpha} \otimes X'^{q,\alpha}$ in $C \circ C'$ is denoted by $\Sigma_\alpha X^{p,\alpha} \circ X'^{q,\alpha}$; the operation $\circ$ is commutative, associative and distributive for the addition.

Example. If C and C' have bases, $\{X^{p,\alpha}\}$ and $\{X'^{q,\beta}\}$, then $C \circ C'$ has the base $\{X^{p,\alpha} \circ X'^{q,\beta}\}$, where $X^{p,\alpha}$ and $X'^{q,\beta}$ are such that:

$$S(X^{p,\alpha} \circ X'^{q,\beta}) = S(X^{p,\alpha}) \cap S(X'^{q,\beta}) \neq \emptyset .$$

Example. If C is fine, then $C \circ C'$ is fine, but in general $\xi^*(C)$ is not fine.

2. THE COHOMOLOGY OF E.

The cohomology of a complex C. The elements X of C such that dX = 0 are called cocycles of C; they constitute a subring Z(C) of C. For instance, the unity U of C is a cocycle. The image dC of C by d is an ideal dC of Z(C); any X belonging to dC is said to be cohomologous to zero, which is written:

$$X \sim 0.$$

The quotient ring

$$H(C) = Z(C)/dC$$

is called the cohomology of C; its elements are the classes of cohomologous cocycles.

H(C) is *graded* and has a *unity*: the class of U.

H(C) is said to be *trivial* when it has a base reduced to its unity.

Define a *coverture* (i.e.: a simple grating in [1]) of E as a complex K of E such that $H(K \cdot x)$ is trivial $\forall x \in E$. (Hence: $K \cdot x$ has elements $\neq 0$, $\forall x \in E$).

The *Čech-cohomology of a compact space* E can be defined as follows (without nerves of covers nor limits): each compact space E has fine covertures; if K and K′ are two of them, then there is a natural isomorphism

$$H(K) \Longleftrightarrow H(K') \qquad (1)$$

which enables us to identify H(K) with H(K′) and to define as follows the cohomology H(E) of E:

$$H(E) = H(K) .$$

The proof of (1) proceeds as follows: the product of K by the unity of K′ defines obviously a natural monomorphism

$$K \subset K \circ K'$$

which induces a natural morphism

$$H(K) \to H(K \circ K') .$$

We assert: "that morphism is an isomorphism"; in other words: "each element X of $K \circ K'$ such that $dX \in K$ satisfies[5] $X \sim K$". Indeed, define the filtration $f(X)$ of $X \in K \circ K'$ as follows: $-f(X)$ is the maximal degree of the elements $X'^{q,\alpha}$ of K' appearing in the expression

$$X = \sum_{\alpha} X^{p,\alpha} \circ X'^{q,\alpha}, \text{ where } p = p(\alpha), \quad q = q(\alpha) ,$$

that expression being chosen such that this maximal degree is as small as possible; m being given, the elements of $K \circ K'$ whose filtrations are $> m$ constitute a subgroup of $K \circ K'$.

Our assumptions

$$f(dX) = 0, \quad K \text{ is fine}, \quad H(K' \cdot x) \text{ is trivial } \forall x \in E$$

imply that:

$X \sim 0$ mod the subgroup of filtration $> f(X)$, if $f(X) < 0$;

$X \in K$ if $f(X) = 0$.

By induction on $f(X)$, our assertion is proved.

The preceding proof belongs to a technique called "filtrated cohomology" and "spectral sequence," the present case being the simple one: the case where the spectral sequence is trivial.

Example. Let K be a (fine or not) coverture of E; then the product of K by the unity of a fine coverture K' of E induces a natural monomorphism

$$K \subset K \circ K'$$

which induces, since $K \circ K'$ is a fine coverture of E, a natural morphism

$$H(K) \to H(E) ;$$

thus each cocycle of K has a natural image in $H(E)$: its cohomology class. A cocycle (or an element) of a coverture of E is called cocycle (or cochain) of E.

5 $X \sim K$ means: there is a $Y \in K \circ K'$ such that $X - dY \in K$.

Example. H(E) is trivial when E is retractible by deformation into a point (for instance: is a cone; is convex).

Example. $H^0(E)$ (i.e.: the subgroup of H of degree 0) is trivial if and only if E is connected.

3. THE COHOMOLOGY OF $E \circ C$.

More generally the proof of (1) gives the following result: Let C be a complex of E, K and K′ be two fine covertures of E; then there is a natural isomorphism

(2) $$H(K \circ C) \Longleftrightarrow H(K' \circ C)$$

which enables us to identity $H(K \circ C)$ with $H(K' \circ C)$ and to define $H(E \circ C)$ by

$$H(E \circ C) = H(K \circ C) .$$

The product of C by the unity of K induces a natural morphism

$$H(C) \to H(E \circ C) .$$

This morphism is an isomorphism, identifying H(C) with $H(E \circ C)$, in the following case:

if C is *basic* and if the supports $S(X^{p,\alpha})$ of its base elements have a *trivial cohomology*, then

(3) $$H(C) = H(E \circ C) .$$

Of course, if C is a coverture of E, then

(4) $$H(E \circ C) = H(E) .$$

Example. If C is a basic coverture of E and if the supports of its base elements have a trivial cohomology, then

$$H(C) = H(E) ;$$

therefore H(E) has a finite base; thus E is a very special space.

The fixed points theory uses still more special spaces.

§2. FIXED POINTS THEORY

4. ADEQUATE COVERTURES.

Notation. Let K be a basic complex; define a local order in its base such that $X^{q+1,\gamma} < X^{q,\beta}$ if and only if $X^{q+1,\gamma}$ appears in $dX^{q,\beta}$; $X^{p,\alpha} < X^{q,\beta}$ if and only if there is an ordered sequence $X^{r,\gamma(r)}$ $(r = p, p-1, \dots, q+1, q)$ from $X^{p,\alpha}$ to $X^{q,\beta}$. Denote by $\underline{X}^{p,\alpha}$ the subcomplex of K generated by the $X^{q,\beta}$ such that $X^{p,\alpha} < X^{q,\beta}$; define

$$S(\underline{X}^{p,\alpha}) = \bigcup_{q,\beta} S(X^{q,\beta}) \text{ for } X^{p,\alpha} < X^{q,\beta} .$$

Definition. An *adequate coverture* K is a basic coverture with the following property:

> if $X^{p,\alpha}$ is a base element, then the subset $S(X^{p,\alpha})$ of E and the subcomplex $\underline{X}^{p,\alpha}$ of K have both trivial cohomology.

The preceding example shows:

If K *is an adequate coverture of* E, *then:*

(5) $$H(E) = H(K);$$

(6) $S(\underline{X}^{p,\alpha})$ has trivial homology, $\forall X^{p,\alpha}$ base element of K.

Definition. The dual k of a base complex K of E is a basic complex of E; its base elements $x_{p,\alpha}$ correspond one-to-one to the base elements $X^{p,\alpha}$ of K; p is called dimension of $x_{p,\alpha}$; the differential of k, (which decreases the dimension by 1, is called boundary and is denoted by ∂), is defined by the condition that

(7) $$\sum_{p,\alpha} X^{p,\alpha} \otimes x_{p,\alpha} \text{ is a } \textit{cocycle} \text{ of } K \otimes k;$$

the cohomology H(k) of k is called the homology of k; its dual is obviously H(K); the supports of the elements of k are defined as follows:

$$S(x_{p,\alpha}) = S(\underline{X}^{p,\alpha});$$

no product is defined in k.

Properties of the dual k *of an adequate coverture* K. In view of (3) and (6):

(8) $$H(E \circ k) = H(k) .$$

Assume E *connected;* then $H^0(E)$ is trivial (no. 2); therefore all the 0-dimensional base elements $x^{0,\alpha}$ of k belong to the same homology class $u \in H(k)$; u is the base of the 0-dimensional homology of k. Thus the homology class of any "0-dimensional" cocycle of $E \circ k$, i.e., of any cocycle

$$\sum_{\alpha} L^{p,\alpha} \circ x_{p.\alpha} \quad (L^{p,\alpha}\text{: cochains of E of degree } p = p(\alpha)) ,$$

is

$$u \cdot K.\,I.\left(\sum_{\alpha} L^{p,\alpha} \circ x_{p,\alpha}\right)$$

where K. I.(...) is an integer, which depends linearly on that cycle: it is its Kronecker index.

For instance (7) implies the existence of cocycles $Z^{q,\beta}$ of K and cycles $z_{q,\beta}$ of k such that

(9) $$\sum_{p,\alpha} X^{p,\alpha} \otimes x_{p,\alpha} \sim \sum_{q,\beta} Z^{q,\beta} \otimes z_{q,\beta} \text{ in } K \otimes k ;$$

their images, in the quotients of the additive groups H(K) and H(k) by their torsion subgroups, are the elements of bases of those quotients; now the Kronecker index defines a duality between those quotients and those bases:

(10) $$K.\,I.(Z^{q,\beta} \circ z_{r,\gamma}) = (-1)^q \text{ if } q = r,\ \beta = \gamma;\ = 0 \text{ otherwise.}$$

5. CONVEXOID SPACE.

Definition. A convexoid space is a space with the following properties:

E is compact and connected;

for any open cover $\cup_\beta 0_\beta = E$ of E, there is a closed and finite cover

$\cup_\gamma F_\gamma = E$ of E such that

1) it is a finer cover (i.e.: each F_γ belongs to some 0_β);
2) any non-empty intersection $F_{\gamma_0} \cap \cdots \cap F_{\gamma_p}$ of its elements has a trivial cohomology.

Examples. Any polyhedron is convexoid. Any compact and convex subset of a locally convex vector space is obviously convexoid. Any sufficiently smooth[6] manifold is convexoid.

Property. A convexoid space E has arbitrarily fine adequate covertures (i.e.: an open cover $\cup_\beta 0_\beta = E$ of E being given, there is an adequate coverture K of E such that the support of any base element of K belongs to some 0_β).

6. LEFSCHETZ NUMBER.

Let E be a convexoid space; let ξ be a mapping:

$$\xi\colon\ E \to E\,.$$

Let K be an adequate coverture of E, k its dual; (7), (9) and (10) show that the integer

$$\mathrm{K.\,I.}\left(\sum_{p,\alpha} \xi^*(x^{p,\alpha}) \circ x_{p,\alpha}\right) = \mathrm{K.\,I.}\left(\sum_{q,\beta} \xi^*(z^{q,\beta}) \circ z_{q,\beta}\right)$$

is the trace[7] of the morphism

$$\xi^*\colon\ H(E) \to H(E)\,, \tag{11}$$

after division of H(E) by its torsion subgroup; it is denoted by $\Lambda_\xi(E)$ and called the *Lefschetz number*[8] of ξ; thus

---

6 It should be sufficient to assume the manifold finite-dimensional and twice differentiable.

7 Obviously: $\xi^*\colon H^p(E) \to H^p(E)$; by definition

$$\mathrm{Trace}\,[\xi^*\colon H(E) \to H(E)] = \Sigma_p (-1)^p\ \mathrm{Trace}\,[\xi^*\colon H^p(E) \to H^p(E)].$$

8 It is a very classical notion, which has been recently renewed by Atiyah and Bott.

(12) $\Lambda_{\xi}(E) = \text{K. I.}\left(\sum_{p,\alpha} \xi^*(X^{p,\alpha}) \circ x_{p,\alpha}\right) = \text{Trace}[\xi\colon H(E) \to H(E)]$.

It is obviously a topological invariant of ξ; it is more precisely an *invariant of the homotopy class of* ξ.

We have

(13) $$\Lambda_{\xi}(E) = 0 \text{ if } \xi \textit{ has no fixed point};$$

indeed we can then choose K so fine that

$$S(\xi^*(X^{p,\alpha})) \cap S(x_{p,\alpha}) = \emptyset , \qquad \forall\, p, \alpha ,$$

which implies

$$\sum_{p,\alpha} \xi^*(X^{p,\alpha}) \circ x_{p,\alpha} = 0 .$$

7. INDEX OF FIXED POINTS

Definition. Let E be a convexoid space, 0 an open subset of E, $\dot{0}$ its boundary, $\overline{0} = 0 \cup \dot{0}$ its closure and ξ a mapping:

$$\xi\colon \overline{0} \to E .$$

Denote by L° a cochain of E (i.e.: an element of a coverture C of E) such that:

L° is homogeneous of degree 0; $S(L^{\circ}) = \overline{0}$; $S(dL^{\circ}) = \dot{0}$;

$L^{\circ} \cdot x$ is the unity of $C \cdot x$ for any $x \in 0$.

Such a cochain exists.

Let K be an adequate coverture of E, k its dual, $\{X^{p,\alpha}\}$ and $\{x_{p,\alpha}\}$ the bases of K and k; consider the element of $C \circ \xi^*(K) \circ k$:

(14) $$\sum_{p,\alpha} L^{\circ} \circ \xi^*(X^{p,\alpha}) \circ x_{p,\alpha} ;$$

its differential is

$$\sum_{p,\alpha} (dL^{\circ}) \circ \xi^*(X^{p,\alpha}) \circ x_{p,\alpha}$$

since $\Sigma_{p,\alpha} X^{p,\alpha} \otimes x_{p,\alpha}$ is a cocycle; assume that ξ has no fixed point on $\dot{0}$; then

$$S(dL^{\circ}) \cap S(\xi^* X^{p,\alpha}) \cap S(x_{p,\alpha}) = \emptyset\,; \qquad \forall\, p, \alpha\,,$$

when K is chosen fine enough; then (14) is a cocycle of $E \circ k$; its Kronecker index is defined; it can be shown that it does not depend on the choices of K and C, provided that K is fine enough. That Kronecker index is called *the index of the fixed points of* ξ *belonging to* 0; it is denoted by

$$\text{(15)} \qquad i_\xi(0) = i_\xi\{x = \xi(x) \in 0\} = \text{K. I.} \left(\sum_{p,\alpha} L^{\circ} \circ \xi^* X^{p,\alpha} \circ x_{p,\alpha} \right) .$$

Properties of the index. $i_\xi(0)$ is defined when ξ has no fixed point on $\dot{0}$. When 0 and ξ are deformed continuously, then $i_\xi(0)$ remains constant as long as it is defined throughout the deformation.

$i_\xi(0) = 1$, when ξ maps $\overline{0}$ into a point of 0.

$i_\xi(0) = 0$, when ξ has no fixed point (in particular when 0 is empty).

$i_\xi(0) = \Sigma_\alpha\, i_\xi(0_\alpha)$ if the 0_α are pairwise disjoint open subsets of 0 and if $\overline{0} - \cup_\alpha 0_\alpha$ contains no fixed points of ξ.

If E' is a convexoid subspace of E and if

$$\xi(\overline{0}) \subset E'$$

then $i_\xi(0)$ does not change when the pair E, 0 is replaced by E', $0' = 0 \cap E'$. If E has trivial cohomology, then $i_\xi(0)$ depends only on the restriction of ξ to $\dot{0}$.

Assume that $E = E' \times E''$; hence $x = (x', x'')$, where $x \in E$, $x' \in E'$, $x'' \in E''$; $\xi(x)$ has the components $\xi'(x) \in E'$, $\xi''(x) \in E''$.

1) $i_\xi(0)$ does not change when ξ'' is replaced by any other mapping $\bar{0} \to E''$ having the same values on the subset of $\bar{0}$ where $x' = \xi'(x)$;

2) Assume: ξ'' independent of x'; $0 = 0' \times D''$ where D'' is a domain (i.e.: is open and connected);

$$x' \neq \xi'(x) \text{ on } \dot{0}' \times \bar{D}''; \quad x'' \neq \xi''(x'') \text{ on } \dot{D}''.$$

Thus the following indices are defined:

$i_\xi(0) = i\{x = \xi(x) \in 0\}$;

$i_{\xi'}(0') = i\{x' = \xi'(x', x'') \in 0'\}$, which is independent of $x'' \in D''$;

$i_{\xi''}(0'') = i\{x'' = \xi''(x'') \in D''\}$.

Now we have:

$$i_\xi(0) = i_{\xi'}(0') \cdot i_{\xi''}(0'') . \tag{16}$$

8. RELATION BETWEEN INDEX AND LEFSCHETZ NUMBER.

We have

$$i_\xi(E) = \Lambda_\xi(E) , \tag{17}$$

since, L° being the unity of a coverture of E when $0 = E$, (12) and (15) are then identical.

This obvious result can be easily extended as follows. Define $\xi^n(\bar{0})$ $(n = 2, 3, \ldots)$ as the image by ξ of $\bar{0} \cap \xi^{n-1}(\bar{0})$; $\xi^n(\bar{0})$ is a decreasing sequence of compact subsets of E; thus

$$f = \lim_{n \to \infty} \xi^n(\bar{G}) = \bigcap_n \xi^n(\bar{G}) \tag{18}$$

is compact (it can be empty);

$$\xi(f) = f .$$

Assume $f \subset 0$; then $i_\xi(0)$ is defined and

$$i_\xi(0) = \Lambda_\xi(f) ; \tag{19}$$

more generally we have then

$$i_\xi(0) = \Lambda_\xi(F) \tag{20}$$

for any compact F such that

$$f \subseteq \xi(F) \subseteq F \subseteq \bar{0} \ . \tag{21}$$

If F is convexoid, then the Lefschetz number $\Lambda_\xi(F)$ is by definition the trace of the morphism

$$\xi^*: \ H(F) \to H(F) \ , \tag{22}$$

where H(F) is replaced by its quotient by its torsion subgroup; that trace is defined since H(F) has then a finite base.

This definition of $\Lambda_\xi(F)$ is equivalent to the following definition, which has to be used when H(F) has not a finite base; $\Lambda_\xi(F)$ is the trace of the morphism (22), where H(F) is replaced by its quotient by the subgroup of all its elements which are mapped into zero by some $m(\xi^*)^p$ (m, p: integers > 0); that quotient has always a finite base under the assumption (21).

9. THE FIXED POINTS OF THE MAPPINGS OF AN ABSOLUTE NEIGHBORHOOD RETRACT (ANR) INTO ITSELF has been studied upon application of the preceding theory: they are obviously fixed points of mappings of convexoid neighborhood of the retract into the retract itself.

10. APPLICATIONS OF THE FIXED POINT THEORY TO NON-LINEAR ELLIPTIC PROBLEM OF VISIK TYPE ARE MADE EASY BY THE MINTY-BROWDER USE OF MONOTONY.

Such problems are reduced to the following one:

There are given a reflexive Banach space V, its dual V′, a mapping

$$A: \ V \times V \to V'$$

and a compact mapping

$$C: \ V \to V.$$

The equation to be studied is:

$$A(u, C(u)) = v'$$

where $u \in V$ is unknown and $v' \in V'$ is given.

Under convenient assumptions (strong monotony and coercivity of $A(u, v)$ relative to u and convenient continuity assumptions) the study of that equation can be reduced to the study of the fixed points of a mapping itself of a compact and convex subset of V.

BIBLIOGRAPHY

[1] D. G. Bourgin, *Modern Algebraic Topology,* New York, Maxmillan (1963).

[2] F. E. Browder, "On the fixed point index for continuous mappings of locally connected spaces," *Summa Brasiliensis Math., 4* (1960), p. 253-293.

[3] A. Deleanu, "Théorie des points fixes: sur les rétractes de voisinage des espaces convexoides," *Bull. Soc. Math. France, 87* (1959) p. 235-243.

[4] A. Granas, "Generalizing the Hopf-Lefschetz fixed-point theorem for non-compact ANR-s" (This Symposium p.

[5] R. Knill, "Some comments on the relationship of the Lefschetz global index and the Leray local index," (this symposium p.).

[6] J. Leray, "Sur la forme des espaces topologiques et sur les points fixes des représentations. Sur les équations et les transformations," *Journ. de Math., 24* (1945) p. 95-248.

[7] ———, "L'anneau spectral et l'anneau filtré d'homologie d'un espace localement compact et d'une application continue," *Journ. de Math., 29* (1950) p. 1-139.

[8] J. Leray, "Théorie des points fixes: indice total et nombre de Lefschetz," *Bull. Soc. Math. France, 87* (1959) p. 221-233.

[9] J. Leray and J. L. Lions, "Quelques résultats de Visik sur les problèmes elliptiques non linéaires par les méthodes de Minty-Browder," *Bull. Soc. Math. France,* 93 (1965) p. 97-107.

[10] J. Leray and J. Schauder, "Topologie et équations fonctionnelles," *Ann. Ecole Norm. Sup., 51* (1934) p. 45-78.

[11] J. G. De Rham, *Variétés Différentiables,* Hermann, 1955.

Note: The assertion of §1 above (Cohomology of compact spaces) are proved in [7], Ch. I, Ch. II (§1-5) and Ch. III ([7] contains more results); analogous but less handy results were established before in [6].

The assertions of §2 above (no. 4-8) (Fixed-point theory) are proved in [6], Ch. II and VI, which have to be simplified as indicated above [8]. An extension of that result is due to R. Knill [5]. The present paper does not quote the wider theory contained in Ch. VII of [6], which has also to be simplified in the same way.

The no. 8 of §2 above (Relation between index and Lefschetz number) gives the results of [8], which have been completed by Browder, Eells and Granas: see [4].

As for no. 9 of §2 above (Fixed points on ANR) see [3].

Finally, for the understanding of no. 10 of §10 above (Elliptic nonlinear problems), see, for instance, [9].

[9] Let us point out that the present paper slightly extends the meaning of "convexoid," which requires the remark b of no. 74, Ch. VI of [6] to be completed (as suggested by D. G. Bourgin).

WEAKLY COMPACT SETS – THEIR TOPOLOGICAL PROPERTIES AND THE BANACH SPACES THEY GENERATE

by

JORAM LINDENSTRAUSS*

§1. *Introduction.*

The purpose of the present paper is to treat some aspects of the theory of weakly compact sets in Banach spaces. We consider the following three closely related topics:

(i) topological properties of weakly compact sets,

(ii) linear topological properties of weakly compact sets, and

(iii) properties of Banach spaces which are generated by weakly compact sets.

Our main interest is investigating which of the results known to hold in the separable case are valid also in the non-separable case.

It is well known that every separable weakly compact set in a Banach space is (in its weak topology) a compact metrizable space and conversely every compact metric space can be embedded as a norm compact (and thus w compact) subset of a separable Banach space. There are some properties of compact metric spaces which are shared by general w compact sets in a Banach space but not by all compact Hausdorff spaces. The well-known Eberlein-Smulýan theorem [17, p. 430] gives a notable example of such a property. We shall present in this paper some more properties of this type. Our discussion is based on a representation theorem for w compact sets which was proved in [1]. Let Γ be a set of indices and let I_γ denote a copy of the interval $[-1, 1]$ for every $\gamma \in \Gamma$. Let $I_\Gamma = \Pi_{\gamma \in \Gamma} I_\gamma$

* The research reported in this document has been sponsored by the Air Force Office of Scientific Research under Grant AF EOAR 66-18 through the European Office of Aerospace Research (OAR) United States Air Force.

be the cartesian product of the intervals I_γ with the usual product topology. Consider closed subsets K of I_Γ which have the property that for every $k \in K$ and every $e > 0$ the set $\{\gamma;\ |k(\gamma)| > e\}$ is finite. Those sets K (with Γ of course arbitrary) are exactly the topological spaces which are homeomorphic to w compact sets in Banach spaces. We call such compact Hausdorff spaces *Eberlein compacts*.

There is another well-known class of compact Hausdorff spaces which includes all the compact metric spaces, and all the members of which enjoy many of the properties of compact metric spaces—namely the class of *dyadic spaces* (cf. [18], [19], [20], [46] and their references). A topological space is called dyadic if it is the continuous image of the product space $D_\Gamma = \Pi_{\gamma \in \Gamma} D_\gamma$ for some set Γ where D_γ is for every γ the space consisting of two points in the discrete topology. The intersection of these two families of compact Hausdorff spaces is exactly the class of compact metric spaces (cf. Corollary 1 of Theorem 3.8). This fact as well as some other facts indicate a certain approach which may be fruitful in organizing the study of non-metrizable compact Hausdorff spaces. We think that such a study should be made "modulo" the class of compact metric spaces. We have not a clear idea of what we mean by "modulo" but for example K and $K \times H$ should be considered as the same object if H is compact metric. In such a study the two classes of Eberlein compacts and dyadic spaces should serve as two "disjoint" basic building blocks for general compact Hausdorff spaces. By the word "disjoint" we have in mind a modification (made necessary by the intention to work "modulo" compact metric spaces) of the notion used by Fürstenberg in another context in [21]. All these remarks concerning the role of Eberlein compacts and dyadic spaces in the general theory of compact Hausdorff spaces are of a very preliminary nature and at the present stage the problem in such a program is to find the suitable definitions. In this paper we restrict ourselves to the study of Eberlein compacts.

We pass now to the second topic mentioned above—linear topological properties of weakly compact convex sets. This topic is considered not only for its own sake. The weakly compact convex subsets of a Banach space form a convenient tool for studying topics (i) and (iii) because of a well-known theorem of Krein-Smulyan [17, p. 434] which asserts that the closed convex hull of a weakly compact set in a Banach space is again weakly compact. An example of a linear topological property of weakly convex sets in Banach spaces which is not valid for a general compact convex set in a locally convex space is the following refinement of the Krein-Milman theorem; A weakly compact convex set in a Banach space is the closed convex hull of its exposed points (cf. Theorem 6.4).

A normed linear space is said to be generated by a weakly compact set (X is WCG, in short) if there is a w compact subset K of X such that X is the closed linear span of K. The notion of a WCG Banach space is a natural common generalization of the notions of a reflexive and of a separable Banach space. The main feature of the class of WCG Banach spaces is the existence of many non-trivial projections on separable subspaces. The projections in a WCG Banach space X are obtained by first constructing a suitable net of operators (usually not even defined on the whole space) and then passing to a limit. The existence of a "big" w compact set guarantees the existence of the limit. There is quite a large freedom in such a construction and thus it is in many instances possible to construct projections having some additional nice properties. This fact enables the extension to WCG Banach spaces of various theorems which are known to hold for separable Banach spaces—e.g. results concerning the existence of "nice" norms (cf. §5) or the existence of quasi-complements (cf. Theorem 2.5). Generally speaking the extensions are made by using transfinite induction.

The main part of this paper is devoted to a survey of results which already appear in the literature. We present also some new results. Results which are stated here without a proof are accompanied by a reference to a paper or a book where the proof may be found. Concerning theorems which are proved in the books [16], [17], [32] or [44] we usually refer to those books instead of the original papers (the references to the original papers may be found in those books).

We would like to emphasize that this paper should not be considered as a survey of the theory of weakly compact sets in Banach spaces. There is a vast literature on weakly compact sets and a survey of all the literature would require much more space. The most notable subjects which are not discussed here are the results centered around the theorems of Eberlein, Smulýan and Krein, criteria for w compactness in special spaces (e.g. spaces of measures or continuous functions) and the recent results of James. The, by now, classical theorems of Eberlein, Smulýan and Krein and their various extensions and generalizations are treated in detail in almost any recent book on functional analysis (see in particular [16], [17] and [32]). The same is ture concerning the most useful criteria for w compactness in the classical Banach spaces. The important results of James ([24], [25] and [26]) are mainly characterizations of w compact sets in which the deep part is the fact that every set which has a certain property (e.g. every linear functional attains its maximum on it) is w compact. Here our main interest is in studying properties of sets of which we know from the outset that they are w compact.

We consider only w compact sets in Banach spaces. It is certain that at least part of the results can be carried over to some more general classes of linear topological spaces and in particular Fréchet spaces. The extensions of the results of Eberlein, Smulýan and Krein to more general linear topological spaces are discussed in detail in [32]. The reader of this discussion will find out that in the problems of this kind the main difficulty and the essential features of the more general case are found already in the Banach space situation. A study of the possible extensions of the theory presented here to some locally convex spaces which are not Banach spaces is planned. However, at the present stage, the study of the open problems in the Banach space case seems to be more important as well as more promising to yield interesting results.

Notations. The notations used here without explanations may be found e.g. in [17]. We consider only spaces over the reals though the same results and proofs are valid in the complex case. By $L_p(\mu)$, $1 \leq p \leq \infty$ we denote the Banach space of (equivalence classes of) measurable functions on some measure space whose p power is integrable (which are essentially bounded if $p = \infty$); μ denotes the measure. If Γ is a set and μ the measure giving to each point the mass 1 we denote $L_p(\mu)$ by $\ell_p(\Gamma)$ (or ℓ_p if Γ is countably infinite). The subspace of $\ell_\infty(\Gamma)$ consisting of those f for which $\{\gamma;\ |f(\gamma)| > \varepsilon\}$ is finite for every $\varepsilon > 0$ is denoted by $c_0(\Gamma)$. If K is a compact Hausdorff space we denote by C(K) the Banach space of continuous real-valued functions on K with the supremum norm. The dimension of a Banach space X is the smallest cardinal m for which there is a subset F of X of cardinality m such that X is the closed linear span of F. The symbol $\sim$ is used for set-theoretic difference while $A - B$ denotes $\{a - b;\ a \in A,\ b \in B\}$. In order to avoid degenerate special cases in some theorems we do not consider the empty set as a topological space, neither a space consisting of 0 alone as a Banach space.

Acknowledgment. I would like to express my great indebtedness to H. H. Corson who originally introduced me to the subject treated here, and with whom I had very helpful discussions concerning the present paper. In particular Proposition 3.4 is due to him.

§2. *Banach spaces generated by weakly compact sets.*

We begin this section with some elementary partial answers to the question "which are the WCG Banach spaces," It is obvious that separable and reflexive spaces are WCG. Typical examples of Banach spaces which are not WCG are ℓ_∞ and $\ell_1(\Gamma)$ for uncountable Γ. In both spaces every w compact set is separable (for ℓ_∞ this is trivial while for $\ell_1(\Gamma)$ this is a result of Phillips (cf. [16, Cor. 2, p. 33])) and since the spaces are not separable they cannot be WCG.

PROPOSITION 2.1. *Let* X *be a Banach space which is generated by a* w *compact set. Let* Y *be a Banach space such that there is a bounded linear operator from* X *into* Y *whose range is dense in* Y. *Then* Y *is generated by a* w *compact set.*

Proof: Obvious.

COROLLARY 1. *Let* X *be a Banach space which is generated by a* w *compact set. Then every quotient space of* X, *and in particular every complemented subspace of* X, *is generated by a* w *compact set.*

COROLLARY 2. *The space* $L_1(\mu)$ *is generated by a* w *compact set if and only if* μ *is* σ*-finite. The space* $L_\infty(\mu)$ *is generated by a* w *compact set if and only if it is finite-dimensional.*

Proof: If μ is not σ-finite there is a complemented subspace of $L_1(\mu)$ which is isometric to $\ell_1(\Gamma)$ for some uncountable Γ. If μ is σ-finite there is a bounded linear operator from the reflexive space $L_2(\mu)$ into $L_1(\mu)$ whose range is dense. Every infinite dimensional $L_\infty(\mu)$ space has a complemented subspace which is isometric to ℓ_∞.

COROLLARY 3. *The space* $c_0(\Gamma)$ *is generated by a weakly compact set for every set* Γ.

Proof: The formal identity map from $\ell_2(\Gamma)$ into $c_0(\Gamma)$ has a dense range.

PROPOSITION 2.2. *Let* X *be a Banach space which is generated by a* w *compact set. Then the dimension of* X *is equal to the smallest cardinality* m *for which there is a total subset of* X^* *of cardinality* m.

(A subset $F \subset X^*$ *is said to be total if* $x^*(x) = 0$ *for every* $x^* \in F$ *implies that* $x = 0$.*)*

Proof: The proposition is obvious if X is finite-dimensional, so we consider from now on only the infinite-dimensional case. It is well known that if $m = \dim X$ then there is a total subset of X^* of cardinality m. Indeed, if $\{x_\alpha\}$ is a dense subset of X and if for every α, $x_\alpha^* \in X^*$ satisfies $\|x_\alpha^*\| = 1$, $x_\alpha^*(x_\alpha) = \|x_\alpha\|$ then the set $\{x_\alpha^*\}$ is a total subset of X^*. Conversely, assume that F is a total subset of X^* of cardinality m and let K be a weakly compact subset of X which generates X. Consider the operator T: $X^* \to C(K)$ defined by $(Tx^*)(k) = x^*(k)$. By our assumption on F the subset TF of C(K) separates the points of K. By the Stone-Weierstrass theorem C(K) has a norm dense subset of cardinality $\leq \mathfrak{m}$. This implies that K has in its weak topology a dense set of cardinality $\leq \mathfrak{m}$ (take e.g. for every f in the dense subset of C(K) a point of K at which f attains its maximum). Since K generates X, it follows that X has a dense subset (in the w topology and thus in the norm topology by [17, Th. 13, p. 422]) of cardinality at most $\mathfrak{m}$

COROLLARY 1. *Let* X *be a separable Banach space. Then* X^* *is generated by a* w *compact set if and only if* X^* *is separable.*

PROPOSITION 2.3. *Let* X *be a Banach space and let* Y *be a closed subspace of* X. *Assume that* Y *is generated by a* w *compact set and that* X/Y *is separable. Then* X *is generated by a* w *compact set.*

Proof: Let T: $X \to X/Y$ be the quotient map. By a theorem of Bartle and Graves [4] (cf. also Michael [41]) there is a norm continuous nonlinear function ϕ: $X/Y \to X$ such that $T\phi$ is the identity map of X/Y. Let K be a weakly compact subset of Y which generates Y and let H be a norm compact subset of X/Y which generates X/Y. It is easily verified that $\phi H + K$ is a w compact subset of X which generates X.

A Banach space X is called quasireflexive if $X^{**}/J(X)$ is finite-dimensional where J: $X \to X^{**}$ is the canonical embedding.

COROLLARY 1. *A quasireflexive space is generated by a* w *compact set.*

Proof: By a result of Civin and Yood [9, Th. 4.6] there is in every quasireflexive space X a reflexive subspace Y such that X/Y is separable.

If $\{X_\gamma\}_{\gamma \in \Gamma}$ is a set of Banach spaces and $1 \leq p \leq \infty$ or $p = 0$ we denote by $(\Sigma \oplus X_\gamma)_p$ the direct sum of the spaces in the $\ell_p(\Gamma)$ sense (or $c_0(\Gamma)$ sense if $p = 0$) as in [16, p. 31].

PROPOSITION 2.4. *Let* $\{X_\gamma\}_{\gamma \in \Gamma}$ *be a set of Banach spaces which are generated by* w *compact sets. Then* $(\Sigma \oplus X_\gamma)_p$ *is generated by a* w *compact set if* $1 < p < \infty$ *or* $p = 0$. *The same is true if* $p = 1$ *and* Γ *is countable.*

Proof: For every γ let K_γ be a w compact subset of X_γ which generates X_γ and for which $\lambda K_\gamma \subset K_\gamma$ whenever $|\lambda| \leq 1$. Then if $1 < p < \infty$, $K = \{f;\ f \in (\Sigma \oplus X_\gamma)_p,\ f(\gamma) \in K_\gamma,\ \|f\| \leq 1\}$ is a weakly compact set which generates $(\Sigma \oplus X_\gamma)_p$, Since the natural identity map from $(\Sigma \oplus X_\gamma)_2$ into $(\Sigma \oplus X_\gamma)_0$ has a dense range we get that $(\Sigma \oplus X_\gamma)_0$ is WCG. Finally if $p = 1$ and we have a sequence $\{X_n\}_{n=1}^\infty$ of WCG Banach spaces, then $K = \{f;\ f \in (\Sigma \otimes X_n)_1,\ f(n) \in K_n,\ \|f(n)\| \leq n^{-2}\}$ is a w compact set which generates $(\Sigma \oplus X_n)_1$.

A similar result for spaces of integrable functions on a measure space taking values in a WCG Banach space can also be easily proved.

The general question of deciding which spaces are WCG is far from solved as the following problems and example indicate.

Problem 1. *Let* X *be a Banach space which is generated by a* w *compact set. Is every closed subspace of* X *generated by a* w *compact set?*

Problem 2. *Is it true that a Banach space is generated by a* w *compact set if and only if it is Lindelöf in its* w *topology?*

It is clear that an affirmative answer to Problem 2 will give also an affirmative answer to Problem 1. In his paper [10] Corson conjectures that the answer to Problem 2 is affirmative. He proves there several results which support this conjecture. The partial results obtained by Corson do not, however, provide an answer to either the "if" or the "only if" part of Problem 2. It is not known, for example, if $L_1(\mu)$ is Lindelöf in its w topology for every finite measure μ.

Problem 3. *Let* X *be a Banach space such that* X^* *is generated by a* w *compact set. Is* X *generated by a* w *compact set?*

In connection with Proposition 2.3 let us remark that it is not true that if X is a Banach space which has a WCG subspace Y for which also X/Y is WCG, then X itself is WCG. In [10, Example 2] Corson considered the following space. Let X be the Banach space of real-valued functions on [0, 1] for which $f(t+0) = f(t)$ for every t (i.e., f is continuous from the right) and $f(t-0)$ exists for every t, with the supremum norm. Let Y be the subspace of X consisting of continuous functions. Y is separable and thus WCG. The space X/Y is isomorphic to $c_0(\Gamma)$ with Γ the set [0, 1] and thus it is also WCG (Corollary 3 to Proposition 2.1). The space X is not WCG since it is not separable and X^* has a countable total subset (the functionals $\phi_r(f) = f(r)$ with r a rational point in [0, 1]).

We now pass to more profound properties of WCG Banach spaces. As mentioned in the introduction, the basic property is the existence of many projections on separable subspaces. A typical result is the following.

THEOREM 2.1. *Let* X *be a Banach space which is generated by a* w *compact set and let* Y *be a separable subspace of* X. *Then there is a separable subspace* Z *of* X *which contains* Y *and a projection of norm* 1 *from* X *into* Z.

This theorem does not characterize WCG Banach spaces X. It is trivial that the theorem holds also for the non-WCG Banach space $\ell_1(\Gamma)$ (Γ arbitrary). The theorem does not, however, hold if $X = \ell_\infty$, since ℓ_∞ has no complemented infinite-dimensional separable subspaces (indeed, the only complemented infinite.dimensional subspaces of ℓ_∞ are those which are isomorphic to ℓ_∞, cf. [38]).

The proof of Theorem 2.1 for reflexive spaces X is given in [35]. Roughly speaking the proof goes as follows. A separable subspace Z_0 of X is constructed so that for every $e > 0$ and every finite-dimensional subspace B of X there is an operator $T_{B,\varepsilon}: B \to Z_0$ such that $\|T_{B,\varepsilon}\| \leq 1+\varepsilon$ and $T_{B,\varepsilon} y = y$ for every $y \in B \cap Y$. Then by using the w compactness of the unit cell of X and Tychonoff's theorem it is shown that the net $\{T_{B,\varepsilon}\}$ has a limiting point T in the topology of pointwise convergence, taking in X the w topology. This limiting point T is a linear operator from X into its separable subspace Z_0 such that $\|T\| = 1$ and $Ty = y$ for $y \in Y$. Having constructed such a T it is easy to construct a suitable subspace Z. In [35] an iterative argument was used. Another approach is to use the ergodic theorem [17, p. 661] which states that the sequence $(I + T + T^2 + \cdots + T^{n-1})/n$ converges to a projection P. It is easily verified that $Z = Px$ has the desired properties.

In the case of WCG Banach spaces which are not reflexive it is not true that every uniformly bounded set of operators is compact in some reasonable topology. Hence we have to be more careful with the choice of the operators $T_{B,\varepsilon}$ in order to insure that the net $\{T_{B,\varepsilon}\}$ have a suitable limiting point T. It turns out that it is more convenient here to work with the normed space generated algebraically by the w compact set and not the whole Banach space. Thus if X is WCG we have by the Krein- Smulýan theorem a convex w compact subset K of X which is symmetric about the origin, is contained in the unit cell and for which $X = \bigcup_{n=1}^{\infty} nK$. We put $X_0 = \bigcup_{n=1}^{\infty} nK$ (clearly $X = X_0$ only if X is reflexive). In X_0 we have two natural norms—the norm $\| \ \|$ induced from X and the norm $||| \ |||$ whose unit cell is K. By our assumption $\|x\| \leq |||x|||$ for every $x \in X_0$. The operators $T_{B,\varepsilon}$ are constructed now for finite-dimensional subspaces B of X_0 so that $\|T_{B,\varepsilon} x\| \leq (1+\varepsilon)\|x\|$ and $|||T_{B,\varepsilon} x||| \leq (1+\varepsilon)|||x|||$ for every $x \in B$. With such operators the limiting procedures described in the preceding paragraph can be carried out. The details of this argument are not so simple and so we do not reproduce them here (they are given in [35] in the reflexive case and in [1] in the general case).

The argument outlined here for WCG Banach spaces gives the following result.

THEOREM 2.2. *Let X_0 be a linear space on which there are two norms $\| \ \|$ and $||| \ |||$ with $\|x\| \leq |||x|||$ for every $x \in X_0$ Assume that $\{x;\ |||x||| \leq 1\}$ is compact in the w topology induced on X_0 by the functionals which are continuous in the $\| \ \|$ norm. Let Y_0 be a $\| \ \|$ separable subspace of X_0.*

Then there is a $\| \ \|$ *separable subspace* Z_0 *of* X_0 *containing* Y_0 *and a projection* P_0 *from* X_0 *onto* Z_0 *such that* $\|P_0x\| \leq \|x\|$ *and* $|||P_0x||| \leq |||x|||$ *for every* $x \in X_0$.

Let us see how Theorem 2.1 can be deduced from Theorem 2.2. Let X and Y be as in Theorem 2.1 and let K_0 be a symmetric w compact convex subset of the unit cell of X such that $X_0 = \bigcup_{n=1}^{\infty} nK_0$ is dense in X. Without loss of generality we may assume that

(2.1) $$Y_0 = Y \cap X_0 \text{ is norm dense in } Y.$$

Indeed, let K_1 be a norm compact convex symmetric subset of the unit cell of the space Y, which generates Y. Put $K = \text{conv}(K_0 \cup K_1)$. Then K is w compact, convex and symmetric and if we replace in the definition of X_0 the set K_0 by K, then (2.1) clearly holds. Taking in X_0 the norm $||| \ |||$ defined by K we can apply Theorem 2.2 and get suitable Z_0 and P_0. Let Z be the closure of Z_0 in X (in the $\| \ \|$ topology) and let P be the extension of P_0 to an operator on X satisfying $\|Px\| \leq \|x\|$ for every $x \in X$. These Z and P have the properties required in Theorem 2.1.

Theorem 2.2 can be used for obtaining a representation of a WCG Banach space X by WCG Banach spaces whose density character (= dimension in the finite-dimensional case) is smaller than that of X.

THEOREM 2.3. *Let* X *be a Banach space which is generated by a* w *compact set. Let* $\mathfrak{m}$ *be the density character of* X *and let* η *be the first ordinal of cardinality* $\mathfrak{m}$. *Then for every ordinal* $\alpha \leq \eta$ *there is a projection operator* P_α *on* X *such that*

(2.2) $$P_\alpha P_\beta = P_\beta P_\alpha = P_\beta \quad \textit{for } \beta < \alpha,$$

(2.3) $$P_1 = 0, \ P_\eta = \textit{identity operator},$$

(2.4) $$\|P_\alpha\| = 1 \quad \textit{for } \alpha > 1,$$

(2.5) *the density character of* $P_\alpha X$ *is at most the cardinality of* α *for infinite* α,

(2.6) *for every* $x \in X$ *the function* $P_\alpha x$ *is a continuous function from the ordinals (in the order topology) to* X *in its norm topology.*

This theorem is proved in [1]. Like Theorem 2.1 it is proved by first considering a dense subspace X_0 which is algebraically generated by a weakly compact convex symmetric subset K. On X_0 projections $P_{\alpha,0}$ are constructed so that besides $\|P_{\alpha,0}\| = 1$ we have also $|||P_{\alpha,0}x||| \leq |||x|||$ for every $x \in X_0$ where $|||\ |||$ is the norm whose unit cell is K.

Obviously, Theorem 2.3 can be used for extending by transfinite induction some results which are known to hold for separable spaces to general WCG Banach spaces. In [1] for example, the following result was deduced from Theorem 2.3.

THEOREM 2.4. *Let* X *be a Banach space which is generated by a* w *compact set.* Then *there is a set* Γ *and a one-to-one linear operator from* X *into* $c_0(\Gamma)$.

Theorems 2.3 and 2.4 do not characterize WCG Banach spaces. If $X = \ell_1(\Gamma)$, then both theorems are valid for X. If $X = \ell_\infty$ then, as already remarked above, Theorem 2.3 fails to hold for X while Theorem 2.4 is obviously valid for X. If $X = \ell_\infty(\Gamma)$ for uncountable Γ then by a result of Day [15] (cf. also §5 below) Theorem 2.4 also fails to hold for X.

A variant of Theorem 2.3 can be used to get a result concerning quasicomplements. Let X be a Banach space and let Y be a closed subspace of X. A closed subspace Z of X is said to be a quasicomplement of Y if $Y \cap Z = \{0\}$ and $Y + Z$ is dense in X. Mackey [40] (cf. also [23]) proved that in a separable Banach space every closed subspace has a quasicomplement. The following theorem generalizes this result.

THEOREM 2.5. *Let* X *be a Banach space which is generated by a* w *compact set and let* Y *be a closed subspace of* X *which is also generated by a* w *compact set. Then* Y *has a quasicomplement in* X.

The variant of Theorem 2.3 needed for proving Theorem 2.5 is that the projections P_α constructed there have to satisfy besides (2.2)-(2.6) also that $P_\alpha Y \subset Y$ for every α. If X is reflexive the possibility of constructing such P_α, as well as Theorem 2.5 itself, are proved in [37]. For a general WCG Banach space the variant of Theorem 2.3 as well as Theorem 2.5 can be proved by combining the arguments presented in [37] and [1]. We omit the details. Let us just mention that by using the results of Gurarii and Kadec [2] for separable Banach spaces some stronger versions of Theorem 2.5 may be obtained.

§3. *Eberlein compacts.*

A compact Hausdorff space is called an Eberlein compact if it is homeomorphic to a w compact subset of the Banach space $c_0(\Gamma)$ for some Γ. Since a w compact set is norm bounded and since on norm bounded sets the w topology in $c_0(\Gamma)$ coincides with the topology of pointwise convergence on Γ it is clear that the definition given here of an Eberlein compact coincides with the definition given in the introduction.

We begin with some simple observations concerning Eberlein compacts.

PROPOSITION 3.1. *The one point compactification of the disjoint union of Eberlein compacts is again an Eberlein compact.*

Proof: Let $\{K_\alpha\}_{\alpha \in A}$ be a set of Eberlein compacts. Each K_α can be considered as a w compact subset of the unit cell of $c_0(\Gamma_\alpha)$ for some Γ_α with $0 \notin K_\alpha$. Let $\Gamma = \bigcup_{\alpha \in A} \Gamma_\alpha$ and let $\tilde{K}_\alpha$ be the extensions of the elements of K_α to functions on Γ by taking as 0 the coordinates outside Γ_α. Then $\{0\} \cup \bigcup_\alpha \tilde{K}_\alpha$ is a w compact subset of $c_0(\Gamma)$ which is homeomorphic to the one point compactification of the disjoint union of the K_α.

PROPOSITION 3.2. *The one point compactification of a locally compact metric space is an Eberlein compact.*

Proof: It is well known (cf. e.g., [10, p. 2]) that if K is locally compact and metrizable then K is the disjoint union of a set of locally compact separable metric spaces $\{K_\alpha\}$. For every α the one point compactification $K_\alpha \cup \{\infty\}$ of K_α is a compact metric space and thus can be mapped homeomorphically by a map ϕ into the unit cell of $c_0(N_\alpha)$ (with N_α a copy of the integers) so that $\phi\{\infty\} = 0$. Constructing $\tilde{\phi}(K_\alpha)$ as in Proposition 3.1 we get that $\{0\} \cup \bigcup_\alpha \tilde{\phi}(K_\alpha)$ is a w compact subset of $c_0(N_\alpha)$ which is homeomorphic to the one-point compactification of K.

PROPOSITION 3.3. *The cartesian product of a set* $\{K_\alpha\}$ *of Eberlein compacts is an Eberlein compact if and only if at most a countable number of the factors are non-trivial.*

A factor is called trivial if it consists of a single point.

Proof: Clearly a closed subset of an Eberlein compact is again an Eberlein compact. The product of an uncountable number of copies of the space consisting of two discrete points is not an Eberlein compact (use e.g., the fact that there are sequences in such a product which do not have a convergent subsequence). This proves one part of the proposition. To prove the other part let $\{K_j\}_{j=1}^{\infty}$ be a sequence of non-trivial Eberlein compacts. We may assume that K_j is a w compact subset of the cell with radius $1/j$ around 0 in $c_0(\Gamma_j)$. For every integer n map $\Pi_{j=1}^{\infty} K_j$ into $c_0(\Gamma_1 \times \Gamma_2 \times \cdots \times \Gamma_n)$ by putting $\phi_n(k)\,(\gamma_1, \gamma_2, \ldots, \gamma_n) = \Pi_{j=1}^{n} k_j(\gamma_j)$ if $k = (k_1, k_2, \ldots)$. Let

$$\Gamma = \bigcup_{n=1}^{\infty} \prod_{j=1}^{n} \Gamma_j \, .$$

The map $\phi\colon \Pi_{j=1}^{\infty} K_j \to c_0(\Gamma)$ defined by $\phi(k) = (\phi_1(k), \phi_2(k), \ldots)$ is a one-to-one and continuous map (taking in $c_0(\Gamma)$ the w topology). This concludes the proof.

An Eberlein compact was defined as a w compact set in the particular Banach space $c_0(\Gamma)$. The importance of Eberlein compacts in the general theory of Banach spaces stems from the fact that they are not only examples of w compact sets but constitute all w compact sets.

THEOREM 3.1. *Every* w *compact set in a Banach space is in its* w *topology an Eberlein compact.*

Theorem 3.1 is an immediate consequence of Theorem 2.4.

The next theorem gives a direct connection between the topological study of Eberlein compacts and the theory of WCG Banach spaces.

THEOREM 3.2. *Let* K *be a compact Hausdorff space. The Banach space* C(K) *is generated by a* w *compact set if and only if* K *is an Eberlein compact.*

Theorem 3.2 was proved in [1]. A closely related result is

THEOREM 3.3. *A Banach space* X *is generated by a weakly compact set if and only if the unit cell* S^* *of* X^* *in its* w* *topology is affinely homeomorphic to a* w *compact subset of* $c_0(\Gamma)$ *for some* Γ.

Proof: Let K be a w compact subset of X which generates X. Then $x^* \to x^*(k)$ is a one-to-one linear operator from X^* to C(K) which is easily seen to be continuous if X^* is taken in the w* topology and C(K) in its w topology. By Theorems 3.2 and 2.4 there is a one-to-one bounded linear operator from C(K) into $c_0(\Gamma)$ for some Γ . The composition of these two operators maps homeomorphically and affinely S^* onto a w compact subset of $c_0(\Gamma)$.

To prove the converse we remark first that the assumption on the unit cell of X^* is equivalent to the assumption that there is a one-to-one weakly compact operator $T: X^* \to c_0(\Gamma)$ whose restriction to the unit cell of X^* is continuous if X^* is taken in the w* topology and $c_0(\Gamma)$ in the w topology. For every $y^* \in \ell_1(\Gamma) = c_0(\Gamma)^*$ the functional T^*y^* is w* continuous on any cell in X^* and hence by [17, p. 428, Th. 6] it is w* continuous on X^*. Thus T^* is a weakly compact operator (cf. [17, p. 485, Th. 8]) from $\ell_1(\Gamma)$ into X (we identify X with its canonical image in X^{**}). Since T is one-to-one it follows immediately that the range of T^* is dense in X. Hence T^* maps the unit cell of $\ell_1(\Gamma)$ into a w compact set which generates X.

COROLLARY 1. *A* WCG *Banach space is generated by a subset which is homeomorphic (in its* w *topology) to the one-point compactification of a discrete set.*

Proof: Let X be WCG and let $T^*: \ell_1(\Gamma) \to X$ be as in the second part of the proof of Theorem 3.3. Let $\{e_\gamma\}_{\gamma \epsilon \Gamma}$ be the unit basis vectors in $\ell_1(\Gamma)$. Then $\{T^*e_\gamma\}_{\gamma \epsilon \Gamma} \cup \{0\}$ has the required properties. For any sequence of distinct indices $\{\gamma_i\}_{i=1}^{\infty}$, $T^*e_{\gamma_i}$ tends w to 0 (observe that we do not claim that $T^*e_\alpha \neq T^*e_\beta$ if $\alpha \neq \beta$).

We mention now two open problems which are closely connected with the preceding results.

Problem 4. *Let* K *be an Eberlein compact and let* H *be a continuous image of* K. *Is* H *an Eberlein compact?*

Problem 5. *Let* X *be a Banach space such that the unit cell* S* *of* X* *is an Eberlein compact in its* w* *topology. Is* X *generated by a weakly compact set?*

Problem 1 is equivalent to the "union" of Problems 4 and 5. If the answer to Problem 1 is positive then the answer to Problems 4 and 5 is also also positive. Indeed, for Problem 4 we can then use the following implications; K an Eberlein compact ⟹ C(K) WCG (Theorem 3.2) ⟹ C(H) WCG (since C(H) is a closed subspace of C(K)) ⟹ H is an Eberlein compact (Theorem 3.2). For problem 5 we have to observe that X is a closed subspace of the WCG Banach space C(S*) (use Theorem 3.2). Conversely if the answers to Problem 4 and Problem 5 are positive the following implications show that the answer to Problem 1 is also positive. Let X be WCG and let Y be a closed subspace of X. Then X WCG ⟹ the unit cell of X* is in its w* topology an Eberlein compact (Theorem 3.3) ⟹ the unit cell of Y* is in its w* topology an Eberlein compact (Problem 4) ⟹ Y is WCG (Problem 5).

Another open problem is

Problem 6. *Let* K *be a compact Hausdorff space. Is* C(K) w *Lindelöf if and only if* K *is an Eberlein compact?*

In view of Theorem 3.2 this is a special case of Problem 2. A closely related problem is

Problem 6′. *Let* K *be a compact Hausdorff space. Is* C(K) *Lindelöf in the topology of pointwise convergence if and only if* K *is an Eberlein compact?*

A known partial answer to Problem 6 (and hence also Problem 6′) is the following result of Corson [10].

THEOREM 3.4. *Let* K *be the one point compactification of a locally compact metric space. Then* C(K) *is Lindelöf in its* w *topology.*

See also the related results [10, Theorems 2 and 3] and [12, Theorem 2.5].

Problem 6 (or 6′) is concerned with the space C(K, R) of continuous functions from K to the metric space R (the reals), or what is the same to the compact metric space [–1, 1]. The "dual" question, for which K is the space C(R, K)–of continuous functions from R to the compact Hausdorff K –a Lindelöf space in the topology of pointwise convergence, was treated in [12]. Here quite a lot is known. The main positive result is

THEOREM 3.5. *Let* H *be a topological space which is the continuous image of a separable metric space. Let* K *be an Eberlein compact. Then* C(H, K) *is a Lindelöf space in the topology of pointwise convergence.*

The proof of this theorem is given in [12]. It is rather long and involved.

Theorem 3.5 was used in [13] for obtaining selection theorems for some set-valued mappings having an Eberlein compact as a range space. Let us first give some definitions. Let H and K be topological spaces and let ϕ be a map from H to the space 2^K of all non-empty subsets of K. The map ϕ is said to be lower semi-continuous if $\{h;\ \phi(h) \cap G \neq \emptyset\}$ is an open subset of H for every open subset G of K. A continuous function $f\colon H \to K$ is said to be a continuous selection of ϕ if $f(h) \in \phi(h)$ for every h.

THEOREM 3.6. *Let* K *be a weakly compact subset of* $c_0(\Gamma)$ *for some set* Γ. *Let* H *be a paracompact topological space such that every* $h \in H$ *has a neighborhood which is the continuous image of a separable metric space (e.g.,* H *may be any locally separable metric space). Let* ϕ *be a lower semi-continuous map* $H \to 2^K$ *such that* $\phi(h)$ *is convex and closed for every* h. *Then* ϕ *admits a continuous selection.*

This theorem is proved in [13]. For metrizable K, Theorem 3.6 reduces to a special case of a selection theorem of Michael [41]. Examples constructed in [12] and [13] show that the assumptions appearing in the statements of Theorems 3.5 and 3.6 cannot be considerably weakened.

As an example of a result which may be deduced from the preceding selection theorem we mention

THEOREM 3.7. *Let* K *be an Eberlein compact and let* H *be a compact metric space. Assume that there is a continuous open map* ψ *from* K *onto* H. *Then there exists a linear operator* T *from* C(K) *into* C(H) *such that*

$$\inf\{f(k);\ k \in \psi^{-1}(h)\} \leq Tf(h) \leq \sup\{f(k);\ k \in \psi^{-1}(h)\} \tag{3.1}$$

for every $h \in H$ *and* $f \in C(K)$.

Theorem 3.7 is proved in [13]. For K metrizable the theorem was first obtained by Michael [42]. Operators which satisfy (3.1) are usually called averaging operators. For further results on averaging operators and their role in the theory of C(K) spaces we refer the reader to paper [46] of Pełczynski where these questions are studied in detail.

The next theorem was proved in [14]. It is a consequence of Theorems 3.2, 3.3, and 6.4 and it is not related to Theorems 3.5-3.7. A point in a topological space is called a G_δ point if it is the intersection of a sequence of open sets. In a compact Hausdorff space a point is a G_δ point if and only if it has a countable basis of neighborhoods. In a metric space every point is a G_δ point.

THEOREM 3.8. *In an Eberlein compact the* G_δ *points form a dense subset.*

COROLLARY 1. *A dyadic Eberlein compact is metrizable.*

Proof: A result of Esenin Volpin (cf. [19, Th. 14]) asserts that a dyadic space which has a dense set of G_δ points is metrizable. The Corollary can be proved in many different ways. We mention here another proof which is closely related to the subject of this paper. In [46, Proposition 8.12] Pełczynski proved that every w compact set in C(K) is separable if K is a dyadic space. This fact together with Theorem 3.2 also proves Corollary 1.

Simple examples can be constructed of non-metrizable Eberlein compacts which satisfy the first axiom of countability (i.e. in which every point is a G_δ). There are even connected examples of this type. Such examples can be obtained by using the "porcupine topology" defined in [6, §VII]. Let $\{K_h\}_{h\in H}$ be a collection of topological spaces so that also the index set H is a topological space. Let $K = \bigcup_{h\in H} K_h$ and let ϕ: K → H be the map which assigns to a point in K_h its index h. Let ψ: H → K be any map such that $\phi\psi$ is the identity of H. In [6] a topology was defined on K which on each K_h coincides with the given topology and such that a net $\{k_\alpha\}$ converges to a point in $K_h \sim \psi(h)$ only if it is eventually in K_h and such that a net $\{k_\alpha\}$ of points which do not belong to K_h converges to $\psi(h)$ if and only if the net $\{\phi(k_\alpha)\}$ converges to h in the topology of H. The following assertions are easily verified. If all the K_h as well as H are Eberlein compacts (resp. connected, satisfy the first axiom of countability) then the same is true for K. We shall prove only the assertion concerning Eberlein compacts. Let H be a w compact subset of $c_0(\Gamma)$ and for every h let K_h be a w compact subset of the unit cell of $c_0(\Gamma_h)$ so that $\psi(h) = 0$. Let $\tilde{\Gamma} = \Gamma \cup \bigcup_{h\in H} \Gamma_h$ and define σ: $K \to c_0(\tilde{\Gamma})$ by $\sigma(k)(\gamma) = \phi(k)(\gamma)$ if $\gamma \in \Gamma$, $\sigma(k)(\gamma) = k(\gamma)$ if $\gamma \in \Gamma_{\phi(k)}$ and $\sigma(k)(\gamma) = 0$ otherwise. It is easy to see that σ is a homeomorphism if $c_0(\tilde{\Gamma})$ is taken in the w topology. Thus if H and K_h, $h \in H$, are all copies of the unit interval then for every choice of ψ the space K is a non-metrizable arcwise connected Eberlein compact which satisfies the first axiom of countability. The following problem seems, however, to be open.

Problem 7. *Let* K *be a* w *compact convex subset of a Banach space in which every point is a* G_δ *point. Is* K *metrizable?*

The answer to Problem 7 is affirmative if K is symmetric. In fact the following slightly stronger result (due to Corson) is true.

PROPOSITION 3.4. *A symmetric convex non-metrizable* w *compact set of a Banach space contains a subset homeomorphic (in its* w *topology) to the one-point compactification of an uncountable set.*

Proof: Let K be a w compact symmetric convex non-metrizable set in a Banach space X. Without loss of generality we may assume that K generates X. Let $m > \aleph_0$ be the density character of X and let η be its initial ordinal. Let $X_0 = \bigcup_{n=1}^{\infty} nK$, let $\| \ \|$ be the given norm in X and let $|||\ |||$ be the norm in X_0 whose unit cell is K. By the results of [1] mentioned in §2 there is a set of projections P_α $1 \le \alpha \le \eta$ defined on X_0 such that (2.2)-(2.6) hold and $P_\alpha K \subset K$ (i.e., $|||P_\alpha||| \le 1$) for every α. It is clear from the construction in [1] that we may assume also that for every $\alpha < \eta$, $(P_{\alpha+1} - P_\alpha)X_0 \neq \{0\}$. Choose for every $\alpha < \eta$ a $y_\alpha \in P_{\alpha+1}K \sim P_\alpha K$ and put $x_\alpha = \frac{1}{2}(y_\alpha - P_\alpha y_\alpha)$. Clearly $0 \neq x_\alpha \in K$, $P_\alpha x_\alpha = 0$ and $P_{\alpha+1}x_\alpha = x_\alpha$. Hence $P_\beta x_\alpha = P_\beta P_\alpha x_\alpha = 0$ if $\beta < \alpha$ and $P_\beta x_\alpha = P_\beta P_{\alpha+1} x_\alpha = P_{\alpha+1}x_\alpha = x_\alpha$ if $\beta > \alpha$. Let $\{x_{\alpha_i}\}_{i=1}^{\infty}$ be a subsequence of the $\{x_\alpha\}_{\alpha<\eta}$ which converges to an element y. We shall prove that $y = 0$ and this will show that $\{0\} \cup \{x_\alpha\}_{\alpha<\eta}$ is the one point compactification of the discrete space $\{x_\alpha\}_{\alpha<\eta}$ of cardinality m. Let $\sigma = \sup\{\alpha_i\}_{i=1}^{\infty}$. Since there is no decreasing sequence of ordinals there is no loss of generality if we assume that $\sigma > \alpha_i$ for every i. Then for $\beta < \sigma$, $P_\beta y = 0$ since $P_\beta x_{\alpha_i} = 0$ for infinitely many i while $P_\sigma y = w - \lim P_\sigma x_{\alpha_i} = w - \lim x_{\alpha_i} = y$. Hence $y = P_\sigma y = \lim P_{\alpha_i} y = 0$ and this concludes the proof.

It is perhaps helpful to consider one example. Let S be the unit cell $\{x;\ \|x\| \le 1\}$ of the reflexive space $\ell_2(\Gamma)$ (with Γ uncountable) taken in the w topology. The G_δ points of S are exactly those points with $\|x\| = 1$. For every point x_0 with $\|x_0\| = t < 1$ the set

$$\{x_0 + te_\gamma\}_{\gamma \in \Gamma} \cup \{x_0\}$$

is homeomorphic to the one point compactification of Γ with x_0 corresponding to the point $\{\infty\}$ (e_γ, $\gamma \in \Gamma$, denotes an orthonormal basis of $\ell_2(\Gamma)$).

To conclude this section we discuss one property with respect to which the general Eberlein compacts behave differently from metric spaces. Let K be a compact Hausdorff space and let H be a closed subset of K. A linear operator T: C(H) → C(K) is called a simultaneous extension operator if for every f ∈ C(H) the restriction of Tf to H is equal to f. A well-known result of Borsuk, Kakutani and Arens asserts that if K ⊃ H with H compact metric, then there is a simultaneous extension operator T: C(H) → C(K) with norm 1 (cf. [46] where an extensive bibliography is given). This result is no longer valid for non-metrizable Eberlein compacts H. A compact Hausdorff space H was called by Pełczynski [46] an L-extensor if for every compact Hausdorff K ⊃ H there is a bounded linear simultaneous extension operator from C(H) to C(K). The L-extensors were studied in detail in [46, §8]. Here we state just the following result.

PROPOSITION 3.5. *An Eberlein compact is an L-extensor if and only if it is metrizable.*

Proof: The "if" part is a weak version of the Borsuk-Kakutani-Arens theorem mentioned above. The "only if" part follows from Theorem 3.2 above and Proposition 8.11 of [46] which asserts that every w compact set in C(H) is metrizable if H is an L-extensor.

The paper [11] is concerned with a study of simultaneous extension operators from C(H) to C(K) with K ⊃ H and H the simplest non-metrizable Eberlein compact–namely the one-point compactification of an uncountable set. From the argument of [11] it follows easily that there are Eberlein compacts K ⊃ H such that there is no bounded linear simultaneous extension operator from C(H) into C(K). (From Proposition 3.5 we can only infer the existence of such a compact Hausdorff K and not a K which is an Eberlein compact.)

§4. *Weakly compact operators and measures on Eberlein compacts.*

The theory of weakly compact operators is clearly closely connected to the theory of weakly compact sets. From the many known facts concerning weakly compact operators (cf. [16] and [17]) we mention just two results which seem to be the closest to our subject.

The first result is due to Dunford and Pettis (cf. [22] and [17, Th. 12, p. 508]).

THEOREM 4.1. *Let* μ *be a measure and let* T *be a* w *compact operator* $T: L_1(\mu) \to X$ *for some Banach space* X. *Then* T *maps every* w *compact set into a norm compact set.*

The second result is due to Grothendieck [22] and Pełczynski [45].

THEOREM 4.2. *Let* K *be a compact Hausdorff space and let* T *be a bounded linear operator from* C(K) *to* X *for some Banach space* X. *Then the following three assertions are equivalent:*

(i) T *is weakly compact;*

(ii) T *maps every weakly compact set into a norm compact set;*

(iii) T *maps every weakly unconditionally convergent series into an unconditionally convergent series.*

Grothendieck applied Theorem 4.1 for obtaining a result concerning the structure of measures on w compact sets.

THEOREM 4.3. *Every finite positive regular measure on a weakly compact set in a Banach space has a separable support.*

The support of a measure is defined as the complement of the set of points which have neighborhoods with measure 0. We shall present here two proofs of Theorem 4.3. The first proof is (essentially) that of Grothendieck. The second proof is based on the structure theorem of w compact sets–Theorem 3.1.

First Proof: Let μ be a Radon probability measure on the w compact set K in the Banach space X. Without loss of generality we may assume that K is convex and symmetric. Consider the operator T: $L_1(\mu) \to X$ defined by $Tf = \int_K xf(x)d\mu(x)$. The integral exists in the weak sense: Tf is the unique element in X such that $x^*(Tf) = \int_K x^*(x)f(x)d\mu(x)$ for every $x^* \in X^*$. It is well known that since K is w compact Tf exists and moreover $Tf \in K$ if $\|f\| \leq 1$. Therefore T is a w compact operator, and hence by Theorem 4.1 it maps every w compact set into a norm compact set. Since $L_1(\mu)$ is WCG (Cor. 2 to Prop. 2.1) it follows that $\overline{TL_1(\mu)}$ is generated by a norm compact set, i.e., it is separable. Let $K_0 = K \cap \overline{TL_1(\mu)}$. K_0 is a closed convex metrizable subset of K. We claim that the support of μ is contained in K_0. Indeed, if $u \notin K_0$ there is an $x^* \in X^*$ and a real t such that $x^*(\mu) > t > \max_{x \in K_0} x^*(x)$. If there is an $f \in L_1(\mu)$ with $f \geq 0$, $\|f\| = 1$ and $f^{-1}(0) \supset \{x;\ x^*(x) \leq t\}$ then $x^*(Tf) \geq t$ and this contradicts the definitions of K_0 and t. Hence $\mu\{x;\ x^*(x) > t\} = 0$ and this concludes the proof.

Second Proof: By Theorem 3.1 we may assume that K is a weakly compact subset of $c_0(\Gamma)$ for some Γ. Let A be the subset of C(K) consisting of the form

$$f(k) = \frac{1}{n}\, \gamma_1(k)\gamma_2(k) \ldots \gamma_n(k) \quad n = 1, 2, \ldots, \quad \{\gamma_i\}_{i=1}^{n} \subset \Gamma\,. \tag{4.1}$$

By the definitions of $c_0(\Gamma)$ and A we have that $f_m \to 0$ pointwise on K for every sequence $\{f_m\}_{m=1}^{\infty}$ of distinct elements in A. By Lebesgue's dominated convergence theorem $\int f_m d\mu \to 0$ for every such sequence. This implies that $\int f\, d\mu \neq 0$ for at most a sequence $\{f_j\}_{j=1}^{\infty}$ of elements in A. Let Γ_0 be the finite or countably infinite set of those elements of Γ which appear in the representation (4.1) of at least one of the f_j. Let $K_0 = \{k$: $k \in K,\ k(\gamma) = 0$ if $\gamma \in \Gamma \sim \Gamma_0\}$. Let g be any element of C(K) which vanishes on K_0. By the Stone-Weierstrass theorem g is a limit (in the norm topology) of linear combinations of elements of the form $k(\gamma_1) \ldots k(\gamma_n)$ with at least one of the γ_i belonging to $\Gamma \sim \Gamma_0$. Hence $\int g\, d\mu = 0$. This proves that the support of μ is contained in the metrizable subset K_0 of K.

Theorem 4.3 might suggest that w compact convex sets behave "well" with respect to Choquet's representation theorem (cf. [44]). This theorem asserts in the separable case that every point k of a convex compact metrizable set K has a representation of the form $k = \int x\, d\mu(x)$ with μ a probability Radon measure with $\mu(K \sim \operatorname{ext} K) = 0$. A similar but more complicated result is valid for a general compact convex set in a locally convex space. Though measures on Eberlein compacts have a separable support it turns out that the pathologies which arise in the non-separable Choquet theorem arise already in the case of Eberlein compacts. For instance in the examples given at the end of [6] the topological spaces can be taken to be Eberlein compacts (cf. section 3 above) and the Banach spaces can be taken to be WCG.

§5. *Strict convexity, smoothness and related properties.*

A Banach space X is called *strictly convex* if $\|x + y\| < \|x\| + \|y\|$ for every $x, y \in X$ which do not lie on the same ray from 0. A Banach space X is called *smooth* if for every $x \in X$ with $\|x\| = 1$ there is only one $x^* \in X^*$ with $x^*(x) = 1 = \|x^*\|$. A real-valued function f on a Banach space is said to be *Gateaux differentiable* at a point x if there is an $x^* \in X^*$ such that for every $y \in X$,

$$f(x + ty) = f(x) + tx^*(y) + g(x, y, t)$$

with $g(x, y, t)/t \to 0$ as $t \to 0$. A Banach space is smooth if and only if the norm is Gateaux differentiable at every $x \neq 0$ (cf. [32, p. 353]). For many purposes it is important to know whether in a given Banach space there is an equivalent smooth or strictly convex norm. The following two theorems seem to include all known positive results in this direction.

THEOREM 5.1. *Let* X *be a Banach space such that there is a one-to-one linear operator* T *from* X *into* $c_0(\Gamma)$ *for some set* Γ. *Then* X *has an equivalent strictly convex norm.*

This theorem is due to Day [15]. Day constructed in that paper a strictly convex norm $|||\ \ |||$ in $c_0(\Gamma)$ which is equivalent to the usual supremum norm. It is obvious that if $\|\ \ \|$ is the given norm in X then $|||x||| = \|x\| + |||Tx|||$ is an equivalently strictly convex norm in X.

THEOREM 5.2. *Let* X_0 *be a subspace of Banach space* X, X *generated by a weakly compact set. Then there is an equivalent strictly convex and smooth norm in* X_0 *such that the norm it induces on* X_0^* *is also strictly convex.*

In [1] it was proved that a WCG Banach space X has an equivalent strictly convex norm $\|\ \ \|_1$ and also an equivalent smooth norm $\|\ \ \|_2$ which induces a strictly convex norm on X^*. (Remarks: by the norm induced by $\|\ \ \|_2$ on X^* we mean $\|x^*\|_2 = \sup\{|x^*(x)|, \|x\|_2 \leq 1\}$. It is trivial that if $\|\ \ \|_2$ on X^* is strictly convex, then $\|\ \ \|_2$ on X is smooth.) By using an ingenious averaging argument of Asplund [2] it follows that in a WCG Banach space X there is one single norm which has the properties of $\|\ \ \|_1$ and $\|\ \ \|_2$ i.e., has all the properties appearing in the statement of Theorem 5.2. It is easily verified by using the Hahn-Banach theorem that the restriction of this norm to X_0 has also all the desired properties. This proves the theorem. For separable spaces, Theorem 5.2 was proved by Klee [30].

The fact that Theorems 5.1 and 5.2 are the strongest known positive results suggests naturally the following problems.

Problem 8. *Let* X *be a strictly convex Banach space. Does there exist a one-to-one linear operator from* X *into* $c_0(\Gamma)$ *for some* Γ?

Day [15] proved that $\ell_\infty(\Gamma)$ (Γ uncountable) does not have an equivalent strictly convex norm.

Problem 9. *Let* X *be a smooth Banach space. Is* X *isomorphic to a subspace of a Banach space generated by a weakly compact set?*

Day proved in [15] that ℓ_∞ and $\ell_1(\Gamma)$ (Γ uncountable) do not have an equivalent smooth norm. A simple proof of these facts was given by Kadec [28]. Kadec used the following proposition which is an immediate consequence of the theorem of Bishop and Phelps [7].

PROPOSITION 5.1. *Let* X *be a smooth Banach space. Then the density character of* X^* *is not greater than the cardinality of* X *(i.e., the "number" of points in* X*).*

Problem 10. *Let* X *be a smooth Banach space. Does* X *have an equivalent strictly convex norm?*

If the answer to Problem 9 is affirmative then by Theorem 5.2 the same is true for Problem 10.

Problem 11. *Characterize the compact Hausdorff spaces* K *such that* C(K) *has an equivalent strictly convex (resp. smooth) norm.*

In recent years it became clear from several investigations (concerning e.g. topological homeomorphisms of Banach spaces or the theory of fixed points; see also the next section) that the following notion introduced by Lovaglia [39] is perhaps even more important than strict convexity. A Banach space X is called *locally uniformly convex* if whenever $\|x\| = \|y_n\| = 1$, $n = 1, 2, \ldots$, and $\|x + y_n\| \to 2$, then $\|x - y_n\| \to 0$. The next problem is of great interest in this connection.

Problem 12. *Let* X *be a Banach space generated by a weakly compact set. Does* X *have an equivalent locally uniformly convex norm? In particular does every reflexive space have an equivalent locally uniformly convex norm?*

The norm of Day [15] in $c_0(\Gamma)$ which was mentioned above is locally uniformly convex. This follows by a straightforward computation. Phelps has remarked (cf. [3]) that the proof of Theorem 5.1 for reflexive X (the operator T exists by Theorem 2.4) gives an equivalent norm on X which is *weakly locally uniformly convex*, i.e., for which $\|x\| = \|y_n\| = 1$, $n = 1, 2, \ldots$, $\|x + y_n\| \to 2$ imply that y_n tends weakly to x. The existence of a one-to-one linear operator from a Banach space X into some $c_0(\Gamma)$ does not in general ensure the existence of an equivalent weakly locally uniformly convex norm in X. This follows from

THEOREM 5.3. *The space ℓ_∞ has no equivalent weakly locally uniformly convex norm.*

Proof: The idea of our proof is taken from Day's proof [15] that $\ell_\infty(\Gamma)$ does not have an equivalent strictly convex norm for uncountable Γ.

Let $\|\ \|$ denote the usual sup norm in ℓ_∞ and let $|||\ |||$ denote an equivalent norm in ℓ_∞ such that $\|x\| \leq |||x|||$ for every x. The support $\sigma(x)$ of $x \in \ell_\infty = \ell_\infty(N)$ (N the integers) is the set $\{i;\ i \in N,\ x(i) \neq 0\}$. Let $\tilde{X}$ be the subset of ℓ_∞ consisting of those x for which $N \sim \sigma(x)$ is infinite. Let $\lambda = \sup\{|||x|||;\ x \in \tilde{X},\ \|x\| = 1\}$ and choose an $x_1 \in \tilde{X}$ with $\|x_1\| = 1$ and $|||x_1||| \geq (3\lambda + 1)/4$. Let N_1 be an infinite subset of $N \sim \sigma(x_1)$ such that $N \sim (\sigma(x_1) \cup N_1)$ is also infinite and let $i_1 \in N \sim (\sigma(x_1) \cup N_1)$. Let

$$F_1 = \{y;\ y \in \ell_\infty,\ \|y\| = 1,\ y(i) = x(i) \text{ for } i \in \sigma(x_1) \cup N_1\ ,$$
$$|y(i_1)| = 1,\ N \sim (\sigma(y) \cup \sigma(x_1) \cup N_1) \text{ is infinite}\}\ .$$

Clearly $2x_1 - F_1 = F_1$. Let

$$m_1 = \inf\{|||y|||;\ y \in F_1\} \text{ and } M_1 = \sup\{|||y|||;\ y \in F_1\}\ .$$

Let $\varepsilon > 0$ and choose a $y \in F_1$ such that $|||y||| \leq m_1 + \varepsilon$. Since $2x_1 - y \in F_1$, we get that $|||2x_1 - y||| \leq M_1$. Hence

$$2|||x_1||| \leq |||y||| + |||2x_1 - y||| \leq m_1 + M_1 + \varepsilon\ .$$

Since e was arbitrary we get that $m_1 \geq 2|||x_1||| - M_1$. Also, since $M_1 \leq \lambda$ and $|||x_1||| \geq (3\lambda + 1)/4$ it follows that $M_1 - m_1 \leq (\lambda - 1)/2$. Let now $x_2 \in F_1$

be such that $|||x_2||| \geq (3M_1 + m_1)/4$ and let N_2 be an infinite subset of $N \sim (\sigma(x_1) \cup N_1 \cup \sigma(x_2))$ such that $N \sim (\sigma(x_1) \cup N_1 \cup \sigma(x_2) \cup N_2)$ is also infinite. Let $i_2 \in N \sim (\sigma(x_1) \cup N_1 \cup \sigma(x_2) \cup N_2)$ and let

$$F_2 = \{y;\ y \in \ell_\infty,\ \|y\| = 1,\ y(i) = x(i) \text{ for } i \in \sigma(x_1) \cup N_1 \cup \sigma(x_2) \cup N_2,$$
$$|y(i_2)| = 1,\ N \sim (\sigma(y) \cup \sigma(x_1) \cup N_1 \cup \sigma(x_2) \cup N_2) \text{ is infinite}\}.$$

Put

$$m_2 = \inf\{|||y|||;\ y \in F_2\} \text{ and } M_2 = \sup\{|||y|||;\ y \in F_2\}\ .$$

A computation similar to the one done above shows that

$$m_2 + M_2 \geq 2|||x_2||| \text{ and } M_2 - m_2 \leq (M_1 - m_1)/2 \leq (\lambda - 1)/2^2\ .$$

Continuing inductively we get a sequence of elements $x_k \in \tilde{X}$, subsets N_k of N, F_k of $\tilde{X}$, real numbers m_k and M_k and integers i_k such that for every k

(5.1) $$x_k \in F_{k-1},\ |||x_k||| \geq (3M_{k-1} + m_{k-1})/4$$

(5.2) N_k is an infinite subset of $N \sim \left(\bigcup_{j=1}^{k-1} N_j \cup \bigcup_{j=1}^{k} \sigma(x_j)\right)$ such that

$$N \sim \left(\bigcup_{j=1}^{k} N_j \cup \bigcup_{j=1}^{k} \sigma(x_j)\right) \text{ is also infinite,}$$

(5.3) $$i_k \in N \sim \left(\bigcup_{j=1}^{k} N_j \cup \bigcup_{j=1}^{k} \sigma(x_j)\right),$$

(5.4) $$F_k = \Big\{y;\ y \in \ell_\infty,\ \|y\| = 1,\ y(i) = x_k(i) \text{ for}$$
$$i \in \bigcup_{j=1}^{k} N_j \cup \bigcup_{j=1}^{k} \sigma(x_j),\ |y(i_k)| = 1,\ N \sim (\sigma(y) \cup \bigcup_{j=1}^{k} N_j \cup \bigcup_{j=1}^{k} \sigma(x_j)) \text{ is infinite}\Big\},$$

(5.5) $$m_k = \inf\{|||y|||;\ y \in F_k\}, \quad M_k = \sup\{|||y|||;\ y \in F_k\},$$

(5.6) $$M_{k-1} \geq M_k \geq m_k \geq m_{k-1}, \quad M_k - m_k \leq (\lambda-1)/2^k .$$

It follows from (5.1) and (5.4) that $x_k(i) = x_{k-1}(i)$ for

$$i \in \bigcup_{j=1}^{k-1} N_j \cup \bigcup_{j=1}^{k-1} \sigma(x_j) .$$

Hence there is an element $x \in \tilde{X}$ such that $x(i) = x_k(i)$ on $\bigcup_{j=1}^{k} N_j \cup \bigcup_{j=1}^{k} \sigma(x_j)$ for every k and $x(i) = 0$ for

$$i \in N \sim \left(\bigcup_{j=1}^{\infty} N_j \cup \bigcup_{j=1}^{\infty} \sigma(x_j) \right)$$

if the set is non-empty. It is clear that such an x belongs to all the sets F_k. By (5.6) there exists $\eta = \lim_{k\to\infty} M_k = \lim_{k\to\infty} m_k$ and hence $|||x||| = \eta = \lim_k |||x_k|||$. Since $(x_k + x)/2 \in F_k$ for every k we get that $|||x + x_k||| \to 2\eta$. If $|||\ \ |||$ were weakly locally uniformly convex this would imply that the sequence x_k tends weakly to x. Let $\theta_k = x(i_k)$, $k = 1, 2, \ldots$. By the definition of x and (5.4) we get that $|\theta_k| = 1$ for every k. Let ϕ be the functional on ℓ_∞ defined by $\phi(y) = \operatorname{LIM}_k (\theta_k y(i_k))$ where LIM denotes a Banach limit (cf. [17, p. 73]). Then, since $x_k(i_j) = 0$ for $j > k$, $\phi(x_k) = 0$ for every k while $\phi(x) = 1$. Thus the sequence $\{x_k\}$ does not tend weakly to x and this proves that $|||\ \ |||$ is not weakly locally uniformly convex. This concludes the proof of the theorem.

For separable Banach spaces X the answer to Problem 12 is affirmative. Kadec proves in [27]

THEOREM 5.4. *In every separable Banach space there is an equivalent locally uniformly convex norm.*

It is of course natural to try to solve Problem 12 by transfinite induction using Theorems 2.3 and 5.4. Although it seems that such an approach will eventually work the difficulties which arise in trying to extend Theorem 5.4 to an arbitrary WCG Banach space (or even only a general reflexive space) have not yet been overcome.

To conclude this section let us mention a problem which is in a sense dual to Problem 12.

Problem 13. *Let* X *be a Banach space such that* X^* *is generated by a weakly compact set. Does* X *have an equivalent Fréchet differentiable norm (at every* $x \neq 0$*)? In particular does every reflexive space have an equivalent Fréchet differentiable norm?*

A real-valued function f defined on a Banach space is said to be *Fréchet differentiable* at a point x if there is an $x^* \in X^*$ such that $f(x+y) = f(x) + x^*(y) + g(x,y)$ with $g(x,y)/\|y\| \to 0$ as $\|y\| \to 0$. A norm in a Banach space is Fréchet differentiable at every point $x \neq 0$ if and only if for every $x \in X$ with $\|x\| = 1$ there is an $x^* \in X^*$ with $x^*(x) = 1 = \|x^*\|$ such that $x^*(y_n) \to 1$. $y_n \in X$, $\|y_n\| = 1 \Longrightarrow \|x - y_n\| \to 0$ (cf. [49]). Again, if X is separable (and thus also X^* is separable by Corollary 1 to Proposition 2.2) the answer to Problem 13 is known to be affirmative (cf. [31] and [47]). It is an immediate consequence of the theorem of Bishop and Phelps [7] that if X has a Fréchet differentiable norm at every point $x \neq 0$ then the density character of X^* is the same as that of X (cf. [28], [47], and [50]. In [50] a much stronger result is proved).

§6. *Extremal structure of convex sets and differentiability of convex functions.*

The results of the preceding sections can be applied to obtain refinements of the well-known Krein-Milman theorem and of Mazur's theorem on the density of smooth points on the unit cell of a separable Banach space (cf. [17, p. 450] and [30]). These refinements will be discussed in the present section. The main tool which enables the application of the results of the preceding sections is

THEOREM 6.1. *Let* X *and* Y *be Banach spaces and let* B(X, Y) *be the Banach space of all bounded linear operators from* X *to* Y. *Let* K *be a* w *compact convex subset of* X *and let*

$$(6.1)\qquad B_K(X, Y) = \{T;\ T \in B(X, Y),\ \textit{there is an}\ x_0 \in K\ \textit{for which}\ \|Tx_0\| = \sup\{\|Tx\|;\ x \in K\}\}.$$

Then $B_K(X, Y)$ *is a norm dense subset of* B(X, Y).

This theorem was proved in [33] for K the unit cell of X (and thus X reflexive). The proof of the general case requires only slight and obvious modifications. Before we turn to the applications of this theorem let us mention two important results which are related to Theorem 6.1 (though there seems to be no direct connection between the three theorems and none can be deduced from the other two).

THEOREM 6.2 *Let* X *be a Banach space and let* K *be a closed bounded convex subset of* X. *Then the set*

$$(6.2)\qquad X_K^* = \{x^*;\ x^* \in X^*,\ \textit{there is an}\ x_0 \in K\ \textit{for which}\ x^*(x_0) = \{\sup x^*(x),\ x \in K\}\},$$

is norm dense in X^*.

Theorem 6.2 is due to Bishop and Phelps [8].

THEOREM 6.3. *Let* X, K *and* X_K^* *be as in Theorem 6.2. Then* $X^* = X_K^*$ *if and only if* K *is weakly compact.*

Theorem 6.3 is due to James [24].

The three preceding theorems should be completed by a fourth theorem asserting when (in the notation of Theorem 6.1) $B(X, Y) = B_K(X, Y)$. A conceivable result of this type is stated as

Problem 14. *Is it true that* $B(X, Y) = B_K(X, Y)$ *if and only if* K *is weakly compact and* $\overline{TK}$ *is norm compact for every* $T \in B(X, Y)$?

It is easy to see that the "if" part is true and that by Theorem 6.3 $B(X, Y) = B_K(X, Y)$ implies that K is weakly compact. Thus the "problem" in Problem 14 is to show that if K is weakly compact and if there is a $T \in B(X, Y)$ with $\overline{TK}$ not norm compact, then $B_K(X, Y) \neq B(X, Y)$:

The relation of Theorem 6.1 to the extremal structure of convex sets is clarified in the next proposition. Let us first introduce some notations. A point x of a convex set K in a Banach space X is called an *exposed point* of K if there is an $x^* \in X^*$ such that $x^*(x) > x^*(y)$ for every $y \neq x$ in K. An exposed point is clearly an extreme point and it is well known that even if $\dim X = 2$ the converse of this statement is false. An exposed point $x \in K$ is said to be a *strongly exposed point* if there is an $x^* \in X^*$ such that $x^*(y_n) \to x^*(x)$, $y_n \in K$, $n = 1, 2, \ldots$, implies that $\|y_n - x\| \to 0$. In a norm compact convex set it is obvious that every exposed point is strongly exposed. Easy examples (cf. [33, p. 145]) show that there exist exposed points of w compact convex sets which are not strongly exposed.

PROPOSITION 6.1. *Let* X *be a Banach space and let* K *be a closed bounded convex and symmetric subset of* X. *Assume that for every Banach space* Y, $B_K(X, Y)$ *is norm dense in* $B(X, Y)$ ($B_K(X, Y)$ *is defined by* (6.1)). *Then*

(i) *if* X *has an equivalent strictly convex norm, then* K *is the closed convex hull of its exposed points;*

(ii) *if* X *has an equivalent locally uniformly convex norm, then* K *is the closed convex hull of its strongly exposed points.*

This proposition is proved in [33] if K is the unit cell of X. The proof of the general case is exactly the same.

As a consequence of Theorem 6.1, Proposition 6.1 and Theorem 5.2 we get the following theorem (first stated in [1], cf. also [14]).

THEOREM 6.4. *A weakly compact convex set in a Banach space is the closed convex hull of its exposed points.*

In the separable case Theorem 6.4 is proved in [29]. Another consequence of Theorem 6.1 and Proposition 6.2 is

THEOREM 6.5. *A weakly compact convex set in a locally uniformly convex Banach space is the closed convex hull of its strongly exposed points.*

COROLLARY 1. *A weakly compact convex set in a separable Banach space is the closed convex hull of its strongly exposed points.*

This corollary is an immediate consequence of Theorem 5.4 and Theorem 6.5, and it was first stated in [33]. A weaker version of Corollary 1 was obtained recently by Namioka [43] by a proof which is very different from the one given in [33] and outlined here. Theorem 6.5 naturally suggests the following problem.

Problem 15. *Is every weakly compact convex set in a Banach space the closed convex hull of its strongly exposed points?*

By Theorem 6.5 a positive answer to Problem 12 will also imply a positive answer to Problem 15. A variant of Problem 15 is formulated in [48]. In this paper a notion of "dentable set" is introduced and applied for obtaining theorems of the Radon-Nikodym type for vector-valued measures. A set which has a strongly exposed point is dentable and it is conjectured in [48] that every w compact set is dentable (cf. [48] and also [43] for details).

Let us now mention another recent refinement of the Krein-Milman theorem. Though on the surface this result does not seem to be related to weakly compact sets, there are strong indications that it is connected to the theory of WCG Banach spaces (see e.g., Problem 16).

THEOREM 6.6. *Let* X *be a separable conjugate Banach space. Then every closed convex and bounded subset of* X *is the norm closed convex hull of its extreme points.*

This theorem is due to Bessaga and Pełczynski [5]. It was first obtained for the special case $X = \ell_1$ in [36] by using a theorem of Bishop and Phelps (Theorem 6.2 above). Bessaga and Pełczynski noted that the proof of [36] for ℓ_1 can be used to prove the general case by using a (non-linear!) homeomorphism between a conjugate separable space X and ℓ_1 which was constructed by Kadec and Klee (cf. [31]). A different proof of Theorem 6.6 was given recently by Namioka [43]. It seems that Theorem 6.6 can be considerably strengthened. The following problem indicates the directions in which this might be possible.

Problem 16. *Let* X *be a conjugate space which is also generated by a* w *compact set. Is every closed convex and bounded subset of* X *the closed convex hull of its extreme (perhaps even exposed or strongly exposed) points?*

We pass now to the dual subject—that of differentiability of convex functions. Here the main two results are

THEOREM 6.7. *Let* X *be a reflexive Banach space which has an equivalent Fréchet differentiable norm at every point* $x \neq 0$ *(in particular* X *may be any separable reflexive space). Then every continuous real-valued convex function* f *defined on an open subset of* X *is Fréchet differentiable in a dense* G_δ *subset of its domain of definition.*

THEOREM 6.8. *Let* X *be a Banach space which is generated by a weakly compact set. Then every continuous real-valued convex function* f *defined on an open subset of* X *is Gateaux differentiable in a dense* G_δ *subset of its domain of definition.*

In the present formulation both theorems are due to Asplund [3]. In [33] it was proved that in a reflexive space X which has a Fréchet differentiable norm every equivalent norm is Fréchet differentiable on a dense subset of X. Only a slight modification of the argument in [33] is needed for proving the same for every convex continuous function defined on an open subset. As shown in [3] it is not hard to deduce the differentiability of a convex continuous function on a dense G_δ set from the differentiability on a dense set. The proof of Theorem 6.7 in [3] is different from that of [33] and it works only under the additional assumption that also X^* has a Fréchet differentiable norm. The proof of Theorem 6.7 in [3] yields, however (by combining it with Theorem 5.2), also a proof of Theorem 6.8. For reflexive Banach spaces X Theorem 6.8 is already a consequence of the results of [33] (only the case $f(x) = ||| \ |||$ where $||| \ |||$ is an equivalent norm is stated explicitly in [33], again without the G_δ insertion).

The fact that the norm is a reflexive Banach space X, which has an equivalent Fréchet differentiable norm, is Fréchet differentiable in a dense subset of X was used in [14] to prove the following representation theorem for weakly compact sets in such spaces.

THEOREM 6.9. *Let X be a reflexive Banach space which has an equivalent Fréchet differentiable norm (in particular X can be any reflexive separable space). Then a subset K of X is weakly compact if and only if it is the intersection of finite unions of cells.*

Remarks added in proof.

1. The answer to problem 7 is negative. Let K be the compact space defined in [14, page 410, remark 2] and let Σ be the set of positive Radon measures μ on K with $\mu(K) = 1$. Then it is not hard to verify that Σ in its w^* topology is a nonmetrizable convex Eberlein compact in which every point is a G_δ point. In view of Proposition 3.4 this set Σ is topologically non-symmetric, i.e., it is not homeomorphic to any convex symmetric Eberlein compact.

2. Concerning the notion of L-extensor which is used in the end of Section 3: In the final version of [46] the term "almost Dugundji space" is used instead of "L-extensor."

REFERENCES

[1] D. Amir and J. Lindenstrauss, "The structure of weakly compact sets in Banach space," to appear.

[2] E. Asplund, "Averaged norms," *Israel J. Math.*, to appear.

[3] ___, "Fréchet differentiability of convex functions," *Acta. Math.*, to appear.

[4] R. G. Bartle and L. M. Graves, "Mappings between function spaces," *Trans. Amer. Math. Soc.*, *72* (1952), 400-413.

[5] C. Bessaga and A. Pełczynski, "Extreme points in separable conjugate spaces," *Israel J. Math.*, *4* (1966).

[6] E. Bishop and K. de Leeuw, "The representation of linear functionals by measures on sets of extreme points," *Ann. Inst. Fourier* (Grenoble) *9* (1959), 305-331.

[7] E. Bishop and R. R. Phelps, "A proof that every Banach space is subreflexive," *Bull. Amer. Math. Soc.*, *67* (1961), 97-98.

[8] _____, "The support functionals of a convex set," *Proc. Symp. Pure Math.*, *7* (1963), 27-35.

[9] P. Civin and B. Yood, "Quasi-reflexive spaces," *Proc. Amer. Math. Soc.*, *8* (1957), 906-911.

[10] H. H. Corson, "The weak topology of a Banach space," *Trans. Amer. Math. Soc.*, *101* (1961), 1-15.

[11] H. H. Corson and J. Lindenstrauss, "On simultaneous extension of continuous functions," *Bull. Amer. Math. Soc.*, *71* (1965), 542-545.

[12] _____, "On function spaces which are Lindelöf spaces," *Trans. Amer. Math. Soc.*, *121* (1966), 476-491.

[13] _____, "Continuous selections with non-metrizable range," *Trans. Amer. Math. Soc.*, *121* (1966), 492-504.

[14] _____, "On weakly compact subsets of Banach spaces," *Proc. Amer. Math. Soc.*, *17* (1966), 407-412.

[15] M. M. Day, "Strict convexity and smoothness of normed spaces," *Trans. Amer. Math. Soc.*, *78* (1955), 516-528.

[16] ———, *Normed Linear Spaces*, Berlin, Springer Verlag, 1958.

[17] N. Dunford and J. T. Schwartz, *Linear Operators, Part I*, Interscience Pub., New York, 1958.

[18] B. Efimov and R. Engelking, "Remarks on dyadic spaces II," *Colloquium Math., 13* (1965), 181-197.

[19] R. Engelking, "Cartesian products and dyadic spaces," *Fundamenta Math., 57* (1965), 287-304.

[20] R. Engelking and A. Pełczynski, "Remarks on dyadic spaces," *Colloquium Math., 11* (1963), 55-63.

[21] H. Fürstenberg, "Disjointness in ergodic theory, minimal sets and a problem in diophantine approximation," *Mathematical Systems Theory, 1* (1967), 1-50.

[22] A. Grothendieck, "Sur les applications linéaires faiblement compactes d'espace du type C(K)," *Canadian J. Math., 5* (1953), 129-173.

[23] V. I. Gurarii and M. I. Kadec, "Minimal systems and quasicomplements in Banach spaces," *Soviet Mathematics, 3* (1962), 966-968 (translated from Russian).

[24] R. C. James, "Weakly compact sets," *Trans. Amer. Math. Soc., 113* (1964), 129-140.

[25] ———, "Weak compactness and separations," *Canadian J. Math., 16* (1964), 204-206.

[26] ———, "Weak compactness and reflexivity," *Israel J. Math., 2* (1964), 101-119.

[27] M. I. Kadec, "Spaces isomorphic to a locally uniformly convex space," *Izv. Vyss. Ucebn. Zaved. Matematika, 13* (1959), 51-57 (Russian).

[28] ———, "Differentiable norms in Banach spaces," *Uspehi Math. Nauk, 20* (1965), 183-187 (Russian).

[29] V. Klee, "Extremal structure of convex sets II," *Math. Z., 69* (1958), 90-104.

[30] ———, "Some new results on smoothness and rotundity in normed linear spaces," *Math. Ann., 139* (1959), 51-63.

[31] V. Klee, "Mappings into normed linear spaces," *Fundamenta Math.*, *49* (1960), 25-34.

[32] G. Köthe, *Topologische Lineare Raume*, Springer Verlag, Berlin 1960.

[33] J. Lindenstrauss, "On operators which attain their norm," *Israel J. Math.*, *1* (1963), 139-148.

[34] _____, "On reflexive spaces having the metric approximation property," *Israel J. Math.*, *3* (1965), 199-204.

[35] _____, "On nonseparable reflexive Banach spaces," *Bull. Amer. Math. Soc.*, *72* (1966), 967-970.

[36] _____, "On extreme points in ℓ_1," *Israel J. Math.*, *4* (1966), 59-61.

[37] _____, "On a theorem of Murray and Mackey," *Anais de Acad. Brasileira Cien* (to appear).

[38] _____, "On complemented subspaces of m," *Israel J. Math.* (to appear).

[39] A. R. Lovaglia, "Locally uniformly convex Banach spaces," *Trans. Amer. Math. Soc.*, *78* (1955), 225-238.

[40] G. W. Mackey, "Note on a theorem of Murray," *Bull. Amer. Math. Soc.*, *52* (1946), 322-325.

[41] E. Michael, "Continuous selections I," *Ann. of Math.*, *63* (1956), 361-382.

[42] _____, "A linear mapping between function spaces," *Proc. Amer. Math. Soc.*, *15* (1964), 415-416.

[43] I. Namioka, "Neighborhoods of extreme points," *Israel J. Math.* (to appear).

[44] R. R. Phelps, *Lectures on Choquet's Theorem*, Van Nostrand, Princeton, 1966.

[45] A. Pełczynski, "Banach spaces on which every unconditionally converging operator is weakly compact," *Bull. Acad. Pol. Sci.*, *12* (1962), 641-648.

[46] _____, "Linear extensions, linear averagings and their application to linear topological classification of spaces of continuous functions," *Rozprawy Matem.* (to appear).

[47] G. Restrepo, "Differentiable norms in Banach spaces," *Bull. Amer. Math. Soc.*, *70* (1964), 413-414.

[48] M. A. Rieffel, "Dentable subsets of Banach spaces with applications to a Radon-Nikodym theorem," *Proc. Conf. on Functional Analysis*, Irvine Calif. (to appear).

[49] V. L. Smulýan, "Sur la structure de la sphere unitaire dans l'espace de Banach," *Math. Sb.* (N. S.), *9* (1941), 545-561.

[50] J. H. M. Whitfield, "Differentiable functions with bounded non-empty support on Banach spaces," *Bull. Amer. Math. Soc.*, *72* (1966), 145-146.

THE HEBREW UNIVERSITY OF JERUSALEM

ON CONTINUITY AND APPROXIMATION QUESTIONS CONCERNING CRITICAL MORSE GROUPS IN HILBERT SPACE

E. H. ROTHE

§1. *Introduction.* In recent years, particularly by the work of Paley and Smale, considerable progress was made in extending the Morse Theory of critical points to Hilbert space. (For a survey and literature see [10, Chapter IV] and [1].)

The present paper deals with continuity and approximation questions concerning the groups attached to critical points which are not necessarily non-degenerate.

Let E be a real Hilbert space, let $\langle x, y\rangle$ denote the scalar product of the elements x and y of E, and let $\|x\|$ denote the non-negative square root of $\langle x, y\rangle$. Let V and V_1 be open balls with the origin as center and radii R and R_1 resp. $(0 < R < R_1)$. Let $i = i(x)$ be a scalar, i.e., a real valued function, defined in V_1. We assume

A) the gradient $g = g(x)$ of i exists;

B) $G(x) = g(x) - x$ is completely continuous and satisfies locally a Lipschitz condition. (We note that assumption B) implies that the Paley-Smale condition stated e.g. in [10, p. 165] is satisfied.)

We recall that a point x is called a critical point of i if it satisfies the condition

$$g(x) = 0 , \tag{1.1}$$

and that a number c is called a critical level of i if there exists a critical point x with $i(x) = c$.

C′) $a < b$ are two real numbers which are non-critical, i.e., which are not critical levels.

We assume moreover

D′) On the boundary $\dot{V}$ of V the gradient field g is strictly exterior, i.e., there exists a constant $\bar{m}$ such that

$$(1.2) \qquad \langle x, g(x) \rangle > \bar{m} > 0 \quad \text{for } x \in \dot{V} .$$

Moreover R_1 is chosen so near to R that there are no critical points in $V_1 - V$. (This choice of R_1 is possible because of assumption B).

Assumption B) also implies that there exists a constant M such that

$$(1.3) \qquad \|g(x)\| < M \quad \text{for } x \in V_1 .$$

We note that conditions A)- E) above are identical with conditions A)- E) stated in [7, p. 239] except for condition C) in [7] which is different from condition C′) above, and for condition D′) above which is a strengthened version of condition D) in [7].

As in [7] we use singular homology theory and denote by $H_q(B, A)$ the q-th homology group of the couple (B, A) where $A \subset B \subset V$, and $q = 0, 1, 2, \ldots$. Moreover for any real number e we use the notations

$$(1.4) \qquad i_e = \{x \in V \mid i(x) < e\}, \quad \bar{i}_e = \{x \in V \mid i(x) \leqq e\} .$$

The following definition is basic; it is legitimate on account of Lemma 2.2.

1.1 *Definition.* Let a, b satisfy C′), and let $c(a, b) = c(a, b;\ i)$ denote the set of critical levels c of i satisfying $a < c < b$. Then $H_q(\bar{i}_b, \bar{i}_a)$ is called the critical group of c(a, b), and the rank M_q of this group is the q-th Morse number of c(a, b).

If c(a, b) consists of a single critical level c then this definition agrees with Definition 4.4 of [7] which in turn agrees with the definition given in [5, Section 7] for the finite dimensional case. Moreover in the case of non-degenerate critical points, the definition of M_q agrees (under certain assumptions) with the definition in terms of the index of the quadratic form in Hilbert space giving the second differential of i as was shown in [7, Section 6]. (We note that the assumption of separability of E made there is superfluous since Lemma 6.4 of [7] holds without that assumption; see e.g. [8, Lemma 8.4].)

In Section 2 of the present paper it is shown that the critical group of c(a, b; i) is isomorphic to the critical group of c(a, b; j) if $|i-j|$ and $\|\text{grad } i - \text{grad } j\|$ are small enough:

(1.5) $$H_q(\bar{i}_b, \bar{i}_a) \approx H_q(\bar{j}_b, \bar{j}_a) \, .$$

(Theorem 2.1.)

Section 3 uses the concept of a "layer scalar" defined in [6, p. 437] (see also Definition 3.1 of the present paper). It is shown that given a scalar i there exists an approximating layer scalar j satisfying the assumptions of Theorem 2.1. Consequently (1.5) holds for such j (Theorem 3.1). It is further shown in Section 3 that to the layer scalar j there corresponds a finite dimensional subspace E^n of E having the following property: if ${}^n j$ denotes the restriction of j to the intersection $V_1 \cap E^n$ then

(1.6) $$H_q(\bar{j}_b, \bar{j}_a) \approx H_q(\overline{{}^n j}_b, \overline{{}^n j}_a)$$

(Lemma 3.3), a fact which together with Theorem 3.1 immediately implies Theorem 3.2.

From (1.5) and (1.6) together with the triangulation properties of finite dimensional differentiable manifolds it is then easily seen that for i smooth enough the critical groups of c(a, b; i) are finitely generated and that they are non-trivial at most for a finite number of q (Theorem 4.1).

Section 5 deals with a generalization to the Hilbert space case of the following result by Morse: Let M_q be the q-th Morse number of the critical set σ belonging to a critical value of the scalar i, and let j be a (good enough) approximation to i having only non-degenerate critical points: then (under certain assumptions) the number of critical points of index q of j neighboring σ is at least M_q ([3, p. 149 and p. 175]; for refinements see [5, p. 29]).

§2. *The continuity of the critical groups.* The aim of this section is the proof of Theorem 2.1 below. We start with two preparatory lemmas.

2.1 *Lemma.* Let $i = i(x)$ satisfy A) and B), and let the real number a be non-critical for x in the closure $\overline{V}$ of V. Then there exists a $\delta > 0$ such that the closed interval $[a-\delta, a+\delta]$ contains no critical value c.

Proof: If the assertion of the lemma were not true there would exist a sequence $\{c_n\}$ of critical values (with $c_n \neq a$) converging to a; and a sequence of points $x_n \in \overline{V}$ such that $i(x_n) = c_n$ and such that (1.1) is satisfied for $x = x_n$. But by a familiar argument assumption B) implies the existence of a subsequence of the x_n converging to some point $x_0 \in \overline{V}$. Then $x = x_0$ would satisfy (1.1) while on the other hand $i(x_0) = a$ in contradiction to the assumption that a is not critical.

If now A), B) and C′) are satisfied we may apply Lemma 2.1 to b as well as to a. It follows that there exist numbers α and β such that $b > \beta > c > \alpha > a$ for all c in c(a, b) (see Definition 1.1).

2.2 *Lemma.* Let A), B), C′) and D′) be satisfied. Let α and β be as in the preceding paragraph. Then

$$H_q(\overline{i}_\beta, \overline{i}_\alpha) \approx H_q(\overline{i}_b, \overline{i}_a). \tag{2.1}$$

Proof: Noting that there exist no critical levels in $[a, \alpha]$ and $[\beta, b]$ one sees easily that the proof given in [7, Theorem 4.2] for the special case that c(a, b) consists of a single point c applies also to the present case.

Lemma 2.2 makes Definition 1.1 legitimate. We are now in a position to state

2.1 *Theorem.* Let the real valued function i satisfy the conditions A)-D) of the introduction. Then there exist two positive numbers ε and ε_1

of the following property: let $j = j(x)$ be a real-valued function satisfying A) and B) (with g replaced by $\gamma = \text{grad } j$) and the inequalities

$$|j(x) - i(x)| < \varepsilon \tag{2.2}$$

$$\|\gamma(x) - g(x)\| < \varepsilon_1 . \tag{2.3}$$

Then C′) and D′) are satisfied with i replaced by j, g by γ, and the constant $\bar{m}$ in (1.2) by $\bar{m}/2$. Moreover (1.5) holds.

Proof: We first determine ε. By Lemma 2.1 we may choose an ε such that the closed intervals

$$\tau_a = [a-\varepsilon,\ a+3\varepsilon], \quad \tau_b = [b-\varepsilon,\ b+2\varepsilon] \tag{2.4}$$

contain no critical levels; we also require that their intersection is empty.

To define ε_1 we note that by continuity of i the set

$$i^{-1}(\tau), \qquad \tau = \tau_a \cup \tau_b \tag{2.5}$$

is closed and therefore has a positive distance from the set Γ of the critical points of i the latter being compact ([7, Lemma 4.2]). Thus there exists a positive constant μ such that $i^{-1}(\tau)$ is contained in the set

$$\Gamma_\mu = \{x \in V \mid \delta(x) \geqq \mu M^{-1}\} \tag{2.6}$$

where $\delta(x)$ denotes the distance of x from Γ and where M is the constant appearing in (1.3). It follows from [7, Lemma 4.6] that there exists a constant m such that

$$\|g(x)\| > m > 0 \quad \text{for } x \in i^{-1}(\tau). \tag{2.7}$$

We now set

$$\varepsilon_1 = \min(m/4,\ \bar{m}/2R) \tag{2.8}$$

where R is the radius of V.

Let then j satisfy the assumptions of our theorem. That C′) is satisfied for j is obvious from (2.7), (2.3) and (2.8). Moreover, the theorem's assertion concerning condition D′) is easily verified by using (1.2), (2.3), (2.8) and Schwarz' inequality.

It remains to prove (1.5). To do this it is sufficient to show that the two pairs occurring in (1.5) are homotopically equivalent (see [2] or the list of topological definitions and theorems, pertinent to our investigation, in [7, Section 2]). In other words, we have to construct continuous maps

$$\phi: (\bar{j}_b, \bar{j}_a) \to (\bar{i}_b, \bar{i}_a) \tag{2.9}$$

$$\psi: (\bar{i}_b, \bar{i}_a) \to (\bar{j}_b, \bar{j}_a) \tag{2.10}$$

such that the composite maps $\psi \circ \phi$ and $\phi \circ \psi$ are homotopic to the identity maps of the respective pairs.

To define ϕ we first define the inclusion map

$$\phi_1: (\bar{j}_b, \bar{j}_a) \to (\bar{i}_{b+\varepsilon}, \bar{i}_{a+\varepsilon}) \;. \tag{2.11}$$

This is possible since the couple at the left member is contained in the couple at the right member as is clear from (2.2). We will now define a map

$$\phi_2: (\bar{i}_{b+\varepsilon}, \bar{i}_{a+\varepsilon}) \to (\bar{i}_b, \bar{i}_a) \tag{2.12}$$

and will then set

$$\phi = \phi_2 \circ \phi_1 \;. \tag{2.13}$$

To define $\phi_2(x_0)$ for $x_0 \in \bar{i}_{b+\varepsilon}$ we will first construct a continuous deformation $\phi_2(x_0, t)$ as follows: since by definition of ε none of the sets

$$S^1 = i^{-1}[b, b+\varepsilon], \quad S^2 = i^{-1}[a+\varepsilon, a+2\varepsilon], \quad S^3 = i^{-1}[a, a+\varepsilon] \tag{2.14}$$

contains a critical point of i, the initial value problem

$$\frac{dx}{dt} = -g(x), \qquad x(x_0, 0) = x_0 \tag{2.15}$$

has for x_0 in one of the sets (2.14) one and only one solution $x = x(x_0, t)$ and this solution is defined and stays in V for $0 \leqq t < \infty$ ([7, Lemma 4.9]). Moreover it follows from [7, Lemma 4.12] that for $t \geqq 2\varepsilon/m^2$

$$i(x(x_0, t)) \leqq \begin{cases} b \text{ for } x_0 \in S^1, \\ a \text{ for } x_0 \in S^2 \cup S^3 \end{cases} \tag{2.16}$$

where m is the constant appearing in (2.7). Since $i(x(x_0, t))$ is decreasing in t ([7, Lemma 4.10]) it follows that there exists a unique $t_0 = t_0(x_0)$ with

$$0 \leqq t_0 \leqq 2\varepsilon/m^2 \tag{2.17}$$

such that

$$i(x(x_0, t)) \begin{cases} > b & \text{for } 0 \leqq t < t_0 \\ \leqq b & \text{for } t_0 \leqq t \end{cases} \quad x_0 \in S^1 \\ \begin{cases} > a & \text{for } 0 \leqq t < t_0 \\ \leqq a & \text{for } t_0 \leqq t \end{cases} \quad x_0 \in S^2 \cup S^3 \tag{2.18}$$

We now define $\phi_2(x_0, t)$ for $x_0 \in \bar{i}_{b+\varepsilon}$, $0 \leqq t \leqq 2\varepsilon/m^2$ as follows:

$$\phi_2(x_0, t) = \begin{cases} x(x_0, t) & \text{for } 0 \leqq t < t_0 \\ x(x_0, t_0) & \text{for } t_0 \leqq t \leqq 2\varepsilon/m^2 \end{cases} \; x_0 \in S^1 \\ \begin{cases} x_0 & \text{for } 0 \leqq t \leqq 2\varepsilon/m^2 \quad x_0 \in i^{-1}[a+2\varepsilon, b] \\ x(x_0, \zeta t) & \text{for } 0 \leqq t < t_0 \\ x(x_0, \zeta t_0) & \text{for } t_0 \leqq t \leqq 2\varepsilon/m^2 \end{cases} \; x_0 \in S^2 \\ \begin{cases} x(x_0, t) & \text{for } 0 \leqq t < t_0 \\ x(x_0, t_0) & \text{for } t_0 \leqq t \leqq 2\varepsilon/m^2 \end{cases} \; x_0 \in S^3 \\ x_0 \quad \text{for } 0 \leqq t \leqq 2\varepsilon/m^2 \quad x_0 \in \bar{i}_a \tag{2.19}$$

with

$$\zeta = \zeta(x_0) = \frac{a + 2\varepsilon - i(x_0)}{\varepsilon} . \tag{2.20}$$

Finally we define

$$\phi_2(x_0) = \phi_2(x_0, 2\varepsilon/m^2), \qquad x_0 \in \bar{i}_{b+\varepsilon} . \tag{2.21}$$

Since as already remarked $\bar{j}_b \in \bar{i}_{b+\varepsilon}$ we may define the restriction $\tilde{\phi}_2$ of ϕ_2 to $\bar{j}_b$. Then from (2.11) and (2.13)

$$\phi(x_0) = \phi_2(x_0) = \tilde{\phi}_2(x_0) \quad \text{for } x_0 \in \bar{j}_b . \tag{2.22}$$

Moreover we have

2.3 *Lemma.* *a)* ϕ_2 maps as indicated in (2.12); if ϕ_2 is restricted to $\bar{i}_b$, then it maps $(\bar{i}_b, \bar{i}_a)$ into itself and is homotopic to the identity mapping of this couple.

β) $\tilde{\phi}_2$ maps $(\bar{j}_b, \bar{j}_a)$ into itself and is homotopic to the identity map of this couple.

Proof: *a)* follows directly from (2.18)-(2.21). To prove β) we denote by $\tilde{\phi}_2(x_0, t)$ the deformation obtained from (2.19) by restricting x_0 to $\bar{j}_b$. It will be sufficient to show that

$$(2.23) \qquad \tilde{\phi}_2(x_0, t) \in \begin{cases} \bar{j}_b & \text{for } x_0 \in \bar{j}_b \\ \bar{j}_a & \text{for } x_0 \in \bar{j}_a \end{cases} \qquad 0 \leqq t \leqq 2\varepsilon/m^2 .$$

Clearly (2.23) will follow once it is proved that

$$(2.24) \qquad \frac{dj(\tilde{\phi}_2(x_0, t)}{dt} \leqq 0 \text{ for } x_0 \in \bar{j}_b. \quad 0 \leqq t \leqq 2\varepsilon/m^2 .$$

Now (2.24) is trivial for $x_0 \notin S^1 \cup S^2 \cup S^3$ since by (2.19) $\tilde{\phi}_2(x_0, t)$ is independent of t. For x_0 in the above union $\tilde{\phi}_2(x_0, t)$ is still constant for $t \geqq t_0$, and we may confine the consideration to positive t-values less than t_0. We consider first an $x_0 \in S^1 \cup S^3$. The left member of (2.24) is then by (2.19) and (2.15)

$$< \text{grad } j, \frac{dx(x_0, t)}{dt} > \; = \; -<\gamma, g> \; = \; -<g, g> + <g-\gamma, g> ,$$

and by Schwarz' inequality this is not greater than

$$-\|g\|^2 + \|g-\gamma\| \; \|g\| \; = \; -\|g\| \, [\|g\| - \|g-\gamma\|] \; .$$

But the right member of this equality is by (2.7), (2.3) and (2.8) not greater than $-\|g\|(m - m/4)$. This proves (2.24) for $x_0 \in S^1 \cup S^3$. If $x_0 \in S^2$ the same reasoning holds except that the argument of g and γ is $x(x_0, \zeta t)$. But it follows from (2.20) and the definition (2.14) of S^2 that $0 \leqq \zeta \leqq 1$. Therefore $0 \leqq t\zeta \leqq t < t_0$ and it follows from (2.18), the definition of S^2 and that of t_0 that $x(x_0, \zeta t) \in i^{-1}(\tau_0) \subset i^{-1}(\tau)$ (cf. (2.4), (2.5)) such that the basic inequality (2.7) still holds for $x = x(x_0, \zeta t)$.

We now turn to the definition of the mapping ψ in (2.10). Corresponding to (2.11) we start by defining the inclusion map

$$\psi_1 : (\bar{i}_b, \bar{i}_a) \to (\bar{j}_{b+\varepsilon}, \bar{j}_{a+\varepsilon}) \tag{2.25}$$

which again is possible because of (2.2), and will define a map

$$\psi_2 : (\bar{j}_{b+\varepsilon}, \bar{j}_{a+\varepsilon}) \to (\bar{j}_b, \bar{j}_a), \tag{2.26}$$

and then set

$$\psi = \psi_2 \circ \psi_1 . \tag{2.27}$$

To define ψ_2 we consider the sets

$$T^1 = j^{-1}[b, b+\varepsilon], \quad T^2 = j^{-1}[a+\varepsilon, a+2\varepsilon], \quad T^3 = j^{-1}[a, a+\varepsilon], \tag{2.28}$$

and note that by (2.2), (2.4) and (2.5) these sets are contained in the set $i^{-1}(\tau)$. Consequently (2.7) holds in these sets. From this, (2.3) and (2.8) we see that

$$\|\gamma(x)\| > m/2 > 0 \quad \text{for } x \in T^1 \cup T^2 \cup T^3 . \tag{2.29}$$

This shows in particular that there are no critical points of j in the union of the T^i. Thus C′) is satisfied for j. So is condition D′) as was already pointed out in the paragraph following (2.8), and since A) and B) are satisfied by assumption all the statements made concerning the solution of the initial value problem (2.15) are true for the solution $\xi = \xi(\xi_0, t)$ of the problem

$$\frac{d\xi}{dt} = -\gamma(x), \quad \xi(\xi_0, 0) = \xi_0, \quad \xi_0 \in T^1 \cup T^2 \cup T^3 \tag{2.30}$$

except that (as comparison of (2.7) with (2.29) shows) m has to be replaced by m/2. Thus (cf. (2.17) and (2.18)) we can assert the existence of a unique $\tau_0 = \tau_0(\xi_0)$ with

$$0 \leqq \tau_0 \leqq 8\varepsilon/m^2$$

such that

$$(2.31)\qquad j(\xi(\xi_0), t)) = \begin{cases} > b \text{ for } 0 \leqq t < \tau_0 \\ \leqq b \text{ for } \tau_0 \leqq t \end{cases} \quad \xi_0 \in T^1 \qquad \begin{cases} > a \text{ for } 0 \leqq t < \tau_0 \\ \leqq a \text{ for } \tau_0 \leqq t \end{cases} \quad \xi_0 \in T^2 \cup T^3 .$$

We now define the deformation $\psi_2(\xi_0, t)$ for $\xi_0 \in \bar{j}_{b+\varepsilon}$, $0 \leqq t \leqq 8\varepsilon/m^2$ as follows:

$$(2.32)\qquad \psi_2(\xi_0, t) = \begin{cases} \left.\begin{array}{ll} \xi(\xi_0, t) & \text{for } 0 \leqq t < \tau_0 \\ \xi(\xi_0, \tau_0) & \text{for } \tau_0 \leqq t \leqq 8\varepsilon/m^2 \end{array}\right\} & \xi_0 \in T^1 \\ \xi_0 \quad \text{for } b \leqq t \leqq 8\varepsilon/m^2 & \xi_0 \in j^{-1}[a + 2\varepsilon, b] \\ \left.\begin{array}{ll} \xi(\xi_0, \eta t) & \text{for } 0 \leqq t < \tau_0 \\ \xi(\xi_0, \eta\tau_0) & \text{for } \tau_0 \leqq t \leqq 8\varepsilon/m^2 \end{array}\right\} & \xi_0 \in T^2 \\ \left.\begin{array}{ll} \xi(\xi_0, t) & \text{for } 0 \leqq t < \tau_0 \\ \xi(\xi_0, \tau_0) & \text{for } \tau_0 \leqq t \leqq 8\varepsilon/m^2 \end{array}\right\} & \xi_0 \in T^3 \\ \xi_0 \quad \text{for } 0 \leqq t \leqq 8\varepsilon/m^2 & \xi_0 \in \bar{j}_a \end{cases}$$

with

$$(2.33)\qquad \eta = \eta(\xi_0) = \frac{a + 2\varepsilon - j(\xi_0)}{\varepsilon} .$$

Finally we define

$$(2.34)\qquad \psi_2(\xi_0) = \psi_2(\xi_0, 8\varepsilon/m^2) , \qquad \xi_0 \in \bar{j}_{b+\varepsilon} .$$

Since $\bar{i}_b \in \bar{j}_{b+\varepsilon}$ we may define the restriction $\tilde{\psi}_2$ of ψ_2 to $\bar{i}_b$. Then from (2.25) and (2.27)

$$(2.35)\qquad \psi(\xi_0) = \psi_2(\xi_0) = \tilde{\psi}_2(\xi_0) , \qquad \xi_0 \in \bar{i}_b .$$

Moreover we have

2.4 *Lemma.* *a)* ψ_2 maps as indicated in (2.26); if ψ_2 is restricted to $\bar{j}_b$, then it maps $(\bar{j}_b, \bar{j}_a)$ into itself and is homotopic to the identity mapping of this couple.

β) $\tilde{\psi}_2$ maps the couple $(\bar{i}_b, \bar{i}_a)$ into itself and is homotopic to the identity map of this couple.

Proof. α) follows directly from (2.31)-(2.34).

The proof of β) is so similar to the proof of the β)-part of Lemma 2.3 that we omit the details. Suffice it to point out that the essential part of the former proof was the estimate of $-\langle \gamma, g \rangle$ along the solutions of (2.15). For the present case the corresponding estimate for $-\langle g, \gamma \rangle$ along solutions of (2.30) is required. This latter estimate may be obtained from the former by interchanging the role of g and γ, and using (2.29) instead of (2.7).

Lemmas 2.3 and 2.4 enable us to finish the proof of Theorem 1.1 by showing that

(2.36) $\psi \circ \phi$ is homotopic to the identity map of the couple $(\bar{j}_b, \bar{j}_a)$

(2.37) $\phi \circ \psi$ is homotopic to the identity map of the couple $(\bar{i}_b, \bar{i}_a)$.

Proof of (2.36). Let $x_0 \in \bar{j}_b$. Then (2.22) holds and it follows from (2.12) and Lemma 2.3 β) that $\phi(x_0) \in \bar{i}_b \cap \bar{j}_b$. Since moreover $\psi_1(\xi) = \xi$ for $\xi \in \bar{i}_b$ (by definition of ψ_1 as the inclusion map (2.25)) we see that $\psi\phi(x_0) = \psi_2\tilde{\phi}_2(x_0)$. Now by Lemma 2.4 α) there exists a deformation $h_1(\xi, t)$ of the couple $(\bar{j}_b, \bar{j}_a)$ such that

$$h_1(\xi, 0) = \psi_2(\xi), \quad h_1(\xi, 1) = \xi \ . \tag{2.38}$$

We extend this deformation to $0 \leqq t \leqq 2$ by setting

$$h_1(\xi, t) = \xi \ \text{for} \ 1 < t \leqq 2 \ . \tag{2.39}$$

Then

$$h_1(\xi, t) \ \epsilon \ \begin{cases} \bar{j}_b & \text{for } \xi \in \bar{j}_b \\ \bar{j}_a & \text{for } \xi \in \bar{j}_a \end{cases} \qquad 0 \leqq t \leqq 2. \tag{2.40}$$

On the other hand there exists by Lemma 2.3 β) a deformation $h_2(x, t)$ of the couple $(\bar{j}_b, \bar{j}_a)$ such that

$$h_2(x_0, 1) = \tilde{\phi}_2(x_0), \ h_2(x_0, 2) = x_0 \ . \tag{2.41}$$

We extend this deformation to $0 \leqq t \leqq 2$ by setting

$$h_2(x_0, t) = \tilde{\phi}_2(x_0) \quad \text{for } 0 \leqq t \leqq 1 \; . \tag{2.42}$$

Then

$$h(x_0, t) \in \begin{cases} \bar{j}_b & \text{for } x_0 \in \bar{j}_b \\ \bar{j}_a & \text{for } x_0 \in \bar{j}_a \end{cases} \qquad 0 \leqq t \leqq 2 \, . \tag{2.43}$$

It is then easily verified that $H(x_0, t) = h_1(h_2(x_0, t), t)$ $(0 \leqq t \leqq 2)$ deforms $\psi \circ \phi$ to the identity map of the couple $(\bar{j}_b, \bar{j}_a)$.

This finishes the proof of (2.36). We omit the proof for (2.37) since it is (in an obvious sense) the "dual" of the proof just given.

§ 3. *The critical groups of approximating layer scalars.* The object of this section has been set forth in the introduction. We start by recalling some pertinent definitions and lemmas, mainly from [6, Section 2].

Let E, V, V_1 be as in the introduction. E^n always denotes an n-dimensional subspace of E. For a fixed E^n the projection of the point $x \in E$ into E^n is denoted by x^n.

3.1 *Definition.* Let $s = s(x)$ be a scalar with domain V_1. Let $S(x)$ be defined by

$$s(x) = \langle x, x\rangle/2 + S(x) \; . \tag{3.1}$$

Then $s(x)$ is called a layer scalar if there exists an E^n such that $S(x) = S(x^n)$. More specifically s is then called a layer scalar with respect to E^n.

Let now $i = i(x)$ be an arbitrary scalar with domain V_1, and let $I(x)$ be defined by

$$i(x) = \langle x, x\rangle/2 + I(x) \; . \tag{3.2}$$

3.2 *Definition.* For a given E^n we set

$$i_n(x) = \langle x, x\rangle/2 + I_n(x) \text{ with } I_n(x) = I(x^n) \; . \tag{3.3}$$

$i_n(x)$ (which is obviously a layer scalar) is called a layer (or projection) approximation of i.

The reason for this terminology is given in the following two lemmas.

3.1 *Lemma.* Let $I(x)$ have a completely continuous gradient $G(x)$ in V_1, and let ε be a given positive number. Then there exists an E^{n_0} such that for $E^n \supset E^{n_0}$

$$|i(x) - i_n(x)| < \varepsilon, \qquad x \in V_1. \tag{3.4}$$

This is Theorem 2.2 of [6].

3.2 *Lemma.* In addition to the assumptions of the preceding lemma let G possess a Fréchet differential which is completely continuous as map of $V_1 \times E$ into E. Let ε_1 be a given positive number. Then there exists an E^{n_0} such that for $E^n \supset E^{n_0}$

$$\|g(x) - g_n(x)\| < \varepsilon_1, \qquad x \in V_1 \tag{3.5}$$

where

$$g_n(x) = x + G_n(x) \tag{3.6}$$

with $G_n(x)$ being the projection $(G(x^n))^n$ of $G(x^n)$ into E^n.

This is Theorem 2.1 of [6].

Remark. $g = \text{grad } i$, and (by [6, Lemma 2.3]) $g_n = \text{grad } i_n$. Moreover: if G satisfies locally a Lipschitz condition so does G_n as follows by obvious estimates from the above definition of G_n.

3.1 *Theorem.* Let i satisfy the assumptions A) - D′) of the introduction and let the assumption of Lemma 3.2 concerning the differential of G be satisfied. Then there exists an E^{n_0} such that for $E^n \supset E^{n_0}$

$$H_q(\bar{i}_b, \bar{i}_a) \approx H_q((\overline{i_n})_b, (\overline{i_n})_a) \tag{3.7}$$

where i_n is the layer approximation (with respect to E^n) as defined in Definition 3.2.

Proof: Let ε and ε_1 be two numbers satisfying (2.4) and (2.8) resp. Then by Lemmas 3.1 and 3.2 there exists an E^{n_0} such that for $E^n \supset E^{n_0}$ the inequalities (3.4) and (3.5) hold simultaneously. This together with the remark following Lemma 3.2 shows that i_n (constructed with such E^n)

satisfies the assumptions made on j in Theorem 2.1. This together with our assumptions on i shows that the conclusion (1.5) of Theorem 2.1 holds with $j = i_n$.

Before stating the next theorem we introduce the following notation: if j is a scalar with domain V_1, then for E^n given ${}^nj$ denotes the restriction of j to $V_1 \cap E^n$.

3.2 *Theorem.* Under the assumptions of Theorem 3.1

$$(3.8) \qquad H_q(\bar{i}_b, \bar{i}_a) \approx H_q((\overline{{}^n i_n})_b, (\overline{{}^n i_n})_a) .$$

Proof. By Theorem 3.1 it will be sufficient to prove the isomorphism of the right members of (3.7) and (3.8), and since i_n is a layer scalar this isomorphism follows from

3.3 *Lemma.* Let $s = \|x\|^2/2 + S(x)$ be a layer scalar with respect to E^n defined in V_1, and let $b > a$ be non-critical values of s. Then

$$(3.9) \qquad H_q(\bar{s}_b, \bar{s}_a) \approx H_q((\overline{{}^n s})_b, (\overline{{}^n s})_a) .$$

Proof: Since ${}^n s$ is a restriction of s we have

$$(3.10) \qquad ({}^n s_b, {}^n s_a) \subset (s_b, s_a) .$$

We consider the deformation

$$(3.11) \qquad x_t = (1-t)x + x^n t , \qquad 0 \leqq t \leqq 1 .$$

Since $\|x^n\| \leqq \|x\|$ it follows from (3.11) that $\|x_t\| \leqq \|x\|$ for $0 \leqq t \leqq 1$. Therefore

$$(3.12) \qquad s(x_t) = \|x_t\|^2/2 + S(x_t) \leqq \|x\|^2/2 + S(x_t) .$$

But by (3.11), $x^n = (x_t)^n$, and since s is a layer map we have $S(x_t) = S((x_t)^n) = S(x^n) = S(x)$ and (3.12) shows that $s(x_t) \leqq \|x\|^2/2 + S(x) = s(x)$. This shows that during the deformation the couple at the right member of (3.10) stays in itself while for $t = 1$ it reduces to the couple at the left of

(3.10) as (3.11) shows. The latter couple is pointwise fixed during the deformation since $x = x^n = x_t$ for $x \in E^n$. These facts imply the isomorphism (3.9).

§4. *The finiteness of the rank of the critical groups.*

4.1 *Theorem.* Under the assumptions of Theorem 3.2 the rank $R_q(i; b, a)$ of the group $H_q(\bar{i}_b, \bar{i}_a)$ is finite. Moreover this group is non-trivial at most for a finite number of values of the integer q.

Proof. The following facts are well known from the investigations of Morse and Morse-v. Schaack (see [4] and the earlier papers quoted there): if E is finite-dimensional, if s is a scalar of class C'' (i.e., is twice continuously differentiable) in V_1, and if $<\text{grad } s, x> > 0$ on the boundary $\dot{V}$ of V_1 then the conclusion of Theorem 4.1 is true with i replaced by s. It follows therefore from (3.8) (which is true for any E^n containing a certain E^{n_0}) that Theorem 4.1 will be proved once it is shown that the assumptions on i of this theorem imply those on s just quoted for $s = {}^n i_n$.

Now (for a proper E^n) i_n satisfies the assumptions on j of Theorem 2.1 (cf. the first paragraph of the proof of Theorem 3.1), and therefore by that theorem satisfies (1.2) with g replaced by grad i_n, and $\bar{m}$ by $\bar{m}/2$.

It remains to prove that ${}^n i_n \in C''$. On account of our assumptions on i (in particular of those stated in Lemma 3.2) this assertion obviously follows from the following two lemmas.

4.1 *Lemma* If $G = \text{grad } I$ has a (Fréchet) differential $dG(x; h) = L(x, h)$ then (for any E^n) the differential of the projection approximation $G_n(x) = (G(x^n))^n$ exists and is given by

$$(4.1) \qquad dG_n(x; h) = L_n(x, h) = (L(x^n, h))^n .$$

Moreover, the second differential of $I_n(x) = I(x^n)$ exists and is given by

$$(4.2) \qquad d^2 I(x; h, k) = <L_n(x, h), k> .$$

Proof: (4.1) is easily verified using the definition of the Fréchet differential and that of $G_n(x)$ (see the last line of Lemma 3.2), and using the facts that projection is linear and that $\|y^n\| \leqq \|y\|$ for any $y \in E$.

(4.2) follows from [5, (2.27)], from (4.1), and from the fact that by [6, Lemma 2.3], $G_n = \text{grad } I_n$.

4.2 *Lemma.* In addition to the assumptions of the preceding lemma we suppose that $L(x, h)$ is continuous in x. Then $L_n(x, h)$ and $d^2I_n(x; h, k)$ are continuous in x.

Proof: On account of (4.2) the second assertion follows from the first, and the first is an obvious consequence of the inequality.

$$\begin{aligned} \|L_n(x+k, h) - L_n(x, h)\| &= \|(L(x^n+k^n, h) - L(x^n))^n\| \\ &\leqq \|L(x^n+k^n, h) - L(x^n, h)\| \;. \end{aligned}$$

§5. The general object of this section was explained in the last paragraph of the introduction. The precise result is set forth in

5.1 *Theorem.* Let the scalar i and the scalar j satisfy the assumptions of Theorem 2.1 such that in particular by that theorem a and b are non-critical for j. In addition we assume the following conditions to be satisfied:

i) the scalar i has exactly one critical value c in (a, b) such that $M_q(i, c)$, the q-th Morse number of i at the level c is the rank of the group $H_q(\bar{i}_b, \bar{i}_a)$ (see Definition 1.1 and the paragraph following that definition);

ii) the scalar j has no more than a finite number of critical values in (a, b)

$$\gamma_1 > \gamma_2 > \cdots > \gamma_p \;. \tag{5.1}$$

(We say $p = 0$ if there are no critical values of j in (a, b));

iii) if σ_v denotes the set of critical points of j at level γ_v then σ_v consists of a finite number of points. Then

(5.2) $$M_q(i, c) \leqq \sum_{v=1}^{p} M_q^{v}(j)$$

where $M_q^v(j)$ denotes the q-th Morse number of j at level γ_v and where the right member is defined to be zero if $p = 0$.

Corollary 1. If $M(i, c) = 0$ for at least one q then $p > 0$.

Corollary 2. We replace assumptions ii) and iii) by the following two: j has only non-degenerate critical points, and in some neighborhood of each critical point the differential $\Lambda(x, h)$ of $-x + \text{grad } j$ is completely continuous in h and continuous in x. Then the number of critical points x of j which have a (Morse-) index q and for which $a < i(x) < b$ is at least equal to $M_q(i, c)$. (For definition of the concepts non-degenerate critical point and index in Hilbert space, see e.g. [7, p. 251] and [8, Definition 8.2].

We start with three lemmas whose proof employs methods used in [3], in [5], and particularly in [11].

5.1 *Lemma.* Let j be a scalar satisfying the conditions A), B), C′), D′) and the assumptions ii) and iii) of Theorem 5.1. Let $\hat{z}_q$ be a non-zero element of $H_q(\bar{j}_b, \bar{j}_a)$, i.e., a class of singular chains $z_q \subset \bar{j}_b$ whose boundary $\partial z_q \subset \bar{j}_a$. (For the formal definition of the symbol $c \subset A$ where c is a singular cycle and A a subset of E see the first paragraph of [2, p. 187]; equivalently we will say c lies on A). Then there exists a number γ in (a, b) such that no $z_q \in \hat{z}_q$ lies on j_γ while there is a $z_q^0 \in \hat{z}_q$ which lies on $\bar{j}_\gamma$.

(We note that in case of an empty $\bar{j}_a$ the assertion of the lemma states that axiom I of Seifert-Threlfall, [11, p. 24], is satisfied.)

Proof of Lemma 5.1. Let $\hat{z}_q$ and z_q be as in the statement of the lemma. We denote by $\mu(z_q)$ the maximum of j on z_q (the maximum exists since a singular chain on E is compact as follows from the definition of such a chain, see [2, p. 186]). Now $\mu(z_q)$ is bounded below as z_q varies over $\hat{z}_q$. Indeed $\mu(z_q) > a$ since otherwise $z_q \subset \bar{j}_a$ such that, against assumption, $\hat{z}_q$ would be the zero element of $H_q(\bar{j}_b, \bar{j}_a)$. Thus

(5.3) $$\gamma = \inf \mu(z_q)$$

exists. We assert next that

(5.4) $$b > \gamma_1 \geqq \gamma \geqq \gamma_p > a .$$

$\gamma_p > a$ is obvious from the definition of γ_p since a is non-critical. Suppose now $\gamma < \gamma_p$. Then there would exist a $\bar{\gamma}$ in $[\gamma, \gamma_p)$ and a $\bar{z}_q \in \hat{z}_q$ with $\mu(\bar{z}_q) = \bar{\gamma}$. Since γ_p is the smallest critical value in (a, b) there is no critical value in $[a, \bar{\gamma}]$. Therefore there exists a deformation deforming the couple $(\bar{j}_{\bar{\gamma}}, \bar{j}_a)$ into the couple $(\bar{j}_a, \bar{j}_a)$ (trivially if $a = \bar{\gamma}$, and by [7, Lemma 4.12 otherwise]). $\bar{z}_q$ would be deformed into a cycle on $\bar{j}_a$ which would imply that $\hat{z}_q$ is the zero element of $H_q(\bar{j}_b, \bar{j}_a)$. This contradiction to our assumptions proves $\gamma \geqq \gamma_p$. The inequality $b > \gamma_1$ is obvious from the definitions. For the proof of $\gamma_1 \geqq \gamma$ we consider any β satisfying

(5.5) $$b > \beta > \gamma_1 .$$

Then $[\beta, b]$ contains no critical level and the couple $(\bar{j}_b, \bar{j}_\beta)$ can be deformed into the couple $(\bar{j}_\beta, \bar{j}_\beta)$; any $z_q \in \hat{z}_q$ will then be deformed into a $z_q^1 \in \hat{z}_q$ on $\bar{j}_\beta$ such that $\mu(z_q^1) \leqq \beta$, and $\gamma \leqq \beta$. Since this is true for all β satisfying (5.5) the desired inequality $\gamma \leqq \gamma_1$ follows.

With γ defined by (5.3) we now proceed to prove the existence of a z_q^0 with the properties asserted in our lemma. Assume there is no such z_q^0. Then by definition of γ and due to the fact that there are at most a finite number of critical values in (a, b) there would then exist a β in (γ, b) such that there is no critical value in $(\gamma, \beta]$ and a $z_q^1 \in \hat{z}_q$ such that $\mu(z_q^1) = \beta$. But on account of assumption iii) a deformation $\delta(x, t)$ deforming $(\bar{j}_\beta, \bar{j}_a)$ into $(\bar{j}_\gamma, \bar{j}_a)$ can be constructed: if γ is non-critical we refer again to [7, Lemma 4.12]. If γ is critical we define δ as follows:

$$\delta(x_0, t) = \begin{cases} x_0 & \text{for } x_0 \in \bar{j}_\gamma, \quad 0 \leqq t \leqq 1 \\ x(t) & \text{for } x_0 \in \bar{j}_\beta - \bar{j}_\gamma, \quad 0 \leqq t < 1 \end{cases}$$

where $x(t)$ is the solution of the initial value problem

$$\frac{dx}{dt} = \frac{(\gamma - i(x_0))\ \mathrm{grad}\ j}{\|\mathrm{grad}\ j\|^2}, \qquad X(0) = x_0 .$$

Finally

$$\delta(x_0, 1) = \lim_{t \to 1^-} x(t) \qquad \text{for } x_0 \in \bar{j}_\beta - \bar{j}_\gamma$$

where the existence of the limit is assured by [7, Theorem 5.1]. δ deforms z_q^1 into a $z_q^0 \in \hat{z}_q$ with $\mu(z_q^0) \leqq \gamma$. But by definition of γ the equality sign must hold. This obviously proves our assertion.

5.2 *Lemma.* The number γ of Lemma 5.1 is a critical value of j. (Cf. [11, p. 24, Satz I].)

Proof: Suppose that γ is not critical. Since there are at most a finite number of critical values in (a, b) there would exist an $\varepsilon > 0$ such that the interval $[\gamma - \varepsilon, \gamma]$ contains no critical values. From this we conclude by an argument used above repeatedly, the existence of a $z_q \in \hat{z}_q$ which lies on $\bar{j}_{\gamma - \varepsilon}$, i.e., for which $\mu(z_q) \leqq \gamma - \varepsilon$, a contradiction to the definition of γ.

5.3 *Lemma.* Let j satisfy the assumptions of Lemma 5.1, and let $M_q^v(j)$ be defined as in the line following (5.2). Then

$$\text{rank } H_q(\bar{j}_b, \bar{j}_a) \leqq \sum_{v=1}^{p} M_q^v(j) . \tag{5.6}$$

Proof: Let us denote by r the number at the left member of (5.6), and let $\hat{z}_q^1, \hat{z}_q^2, \ldots, \hat{z}_q^r$ be r independent elements of $H_q(\bar{j}_b, \bar{j}_a)$. Now to each $\hat{z}_q^\rho$ corresponds a number $\gamma = \gamma(\rho)$ having the properties asserted in Lemma 5.1, and there exists an element $z_q^\rho \in \hat{z}_q^\rho$ with $z_q^\rho \subset \bar{j}_{\gamma(\rho)}$. By Lemma 5.2, $\gamma(\rho)$ is a critical value and therefore equals one of the $\gamma_v (v = 1, 2, \ldots, p)$. If $\gamma(\rho) = \gamma_v$ we say z_q^ρ belongs to the critical value γ_v. Let the notation be such that

$$z_q^1, \ldots, z_q^{r_1} \quad \text{belong to } \gamma_1$$

$$z_q^{r_1+1}, \ldots, z_q^{r_1+r_2} \quad \text{belong to } \gamma_2$$

. . . .

Then $r_1 + r_2 + \cdots + r_p = r$. Let now a_v, b_v $(v = 1, 2, \ldots, p)$ be numbers satisfying

$$b \geqq b_1 > \gamma_1 > a_1 \geqq b_2 > \gamma_2 > a_2 > \cdots > a_{p-1} \geqq b_p > \gamma_p > a_p \geqq a .$$

Then the z_p^ρ may be chosen in such a manner that those belonging to γ_v are independent mod a_v (not only mod a which they are by definition). This may be seen by a construction described in [11, p. 25], and the reader is referred to this reference. Since these cycles lie on $\overline{j}_{\gamma_v} \subset \overline{j}_{b_v}$ it follows that $r_v \leqq \operatorname{rank} H_q(\overline{j}_{b_v}, \overline{j}_{a_v}) = M_q^v(j)$. Adding this inequality over v we obtain (5.6).

Proof of Corollaries 5.1 and 5.2. Corollary 1 is an obvious consequence of (5.2). As to Corollary 2 we note that by [7, Lemma 6.3] and by the reference quoted there, the assumptions of the corollary imply that j has only a finite number of critical points such that the assumptions ii) and iii) of Theorem 5.1 are satisfied. Thus (5.2) holds, and our corollary follows from [7, Theorem 6.2] in conjunction with [9, Theorem 7.2].

REFERENCES

[1]. J. Eells, Jr., "A setting for global analysis," *Bulletin A.M.S. 72* (1966), 751-807.

[2]. Eilenberg and N. Steenrod, *Foundations of Algebraic Topology*, Princeton University Press, 1952.

[3]. M. Morse, "Calculus of Variations in the Large," *A.M.S.*, 1934.

[4]. M. Morse, and G. B. VanSchaack, *Ann. Math. 35* (1934), 545-571.

[5]. E. Pitcher, "Inequalities of critical point theory," *Bull. A.M.S., 64* (1958), 1-30.

[6]. E. H. Rothe, "Leray-Schauder Index and Morse Type Numbers in Hilbert Space, *Ann. Math. 55* (1952), 433-467.

[7]. E. H. Rothe, "Some Remarks on Critical Point Theory in Hilbert Space," *Proc. Symp. on Non-Linear Problems,* U.S. Army Center, University of Wisconsin, April 30-May 2, 1962; University of Wisconsin Press, 1963, 233-256.

[8]. _____, "Critical Point Theory in Hilbert Space under General Boundary Conditions," *J. Math. Analysis and Applications, 11* (1965), 357-409.

[9]. _____, "Some Remarks on Critical Point Theory in Hilbert Space," (Continuation). To appear in the *Journal of Mathematical Analysis and Applications.*

[10]. J. T. Schwartz, "Non-linear Functional Analysis," Lecture Notes, Courant Institute of Mathematical Sciences, New York University, 1963-1964.

[11]. H. Seifert and W. Threlfall, *Variationsrechnung im Grossen* (Theorie von Marston Morse), Leipzig und Berlin, 1938.

THE UNIVERSITY OF MICHIGAN

ON HOMEOMORPHISMS OF CERTAIN INFINITE DIMENSIONAL SPACES

Raymond Y. T. Wong

1. **Isotopy Theorem.** In this paper we concern only separable metric space X, especially when X is the Hilbert cube I^∞, the countably infinite product of open unit-intervals $\dot{I}^\infty$ or the Hilbert space ℓ_2.

Let X^∞ denote the countably infinite product of X by itself. A homeomorphism on X means a homeomorphism of X onto itself. Let $G(X)$ denote the group of homeomorphisms on X. For $f,g \in G(X)$, we write $f \sim g$ if f is isotopic to g, by which we mean there exists a mapping F of $X \times I$ into X such that $F|_{X\times 0} = f$, $F|_{X\times 1} = g$ and each $F|_{X\times t} \in G(X)$. We note immediately that " $\sim$ " is an equivalence relation on $G(X)$.

X is said to satisfy *property* Φ if the following $h \in G(X^\infty)$ is isotopic to the identity mapping on X^∞:

$$h(x_1,x_2,x_3,x_4,\ldots.) = (x_2,x_1,x_3,x_4,\ldots.).$$

THEOREM 1. $G(X^\infty)/\sim$ *is trivial if and only if* X *satisfies property* Φ.

It can be verified rather easily that both I and $\mathring{I}$ satisfy property Φ, therefore we have the following Corollary:

COROLLARY 1. $G(I^\infty)/\sim$, $G(\mathring{I}^\infty)/\sim$ *are both trivial.*

Using the fact $\dot{I}^\infty$ is homeomorphic to ℓ_2 [2], we have

COROLLARY 2. $G(\ell_2)/\sim$ *is trivial.*

Corollary 1, 2 settle a question raised by Klee [9] [10]. The details of proof can be found in the author's [13]. We remark that the technique used in [13] can be employed to prove *Corollary 2* directly [15]. We close this section by stating the following Corollary:

COROLLARY 3. *Let* $X = [-1,1]$, *and let* K *be the subset in* X^∞ *defined by* $x \in K$ *if and only if for all but finitely many* i, $x_i = 0$. *Then* $G(K)/\sim$ *is trivial.*

2. Stable Homeomorphism. A homeomorphism h on X is *stable* (Brown-Gluck) if h can be written as a composition of finitely many homeomorphisms on X each of which is the identity on some open subset of X. The main result in [13] concerning this topic is:

THEOREM 2. *Any homeomorphism on* $\mathring{I}^\infty$ *is stable.*

A similar statement for I^∞ is asserted by R. D. Anderson [3]. We note that *Theorem 2* also implies that any $h \in G(\ell_2)$ is stable. Finally we remark that *Corollary 1* can be obtained quite easily from *Theorem 2* by a method of Alexander.

3. Extending Homeomorphism. The general question is to decide (1) under what condition that an embedding h of a (certain) closed subset K of X into X can be extended to a $\tilde{h} \in G(X)$, (2) what closed subset K of X is "wildly embedded" in X, where "wildly embedded" means the complement of K in X is not simply connected.

In regard to (1), let K_t be the subset of I^∞ defined by $x \in K_t$ iff $x_1 = t$.

THEOREM 3. *A homeomorphism* h *of* $K_t (t = 0$ *or* $\frac{1}{2})$ *into* I^∞ *can be extended to a homeomorphism* $\tilde{h} \in G(I^\infty)$ *if and only if* h *can be extended to a neighborhood of* K_t *onto a neighborhood of* $h(K_t)$.

It is natural to ask the following generalization of $K_{\frac{1}{2}}$: Let H_n be the subset of I^∞ defined by $x \in H_n$ if and only if $x_1 = x_2 = \dots = x_n = \frac{1}{2}$.

Question. If h is a homeomorphism of H_n into I^∞ such that h can be extended to a neighborhood of H_n onto a neighborhood of $h(H_n)$, can h be extended to a $\tilde{h} \epsilon G(I^\infty)$?

For $n = 2$, it is false. A rather simple counter-example can be constructed as follows: Let S_1 be the arc in I^3 such that $x_1 = x_2 = \frac{1}{2}$. Let S_2 be another arc in I^3 such that (1) $\pi_1(I^3 - S_1) \not\approx \pi_1(I^3 - S_2)$ and (2) there is a homeomorphism f of a nbd. of S_1 onto a nbd. of S_2 throwing S_1 onto S_2 (such arc S_2 exists quite obviously). Extend f to $f \times 1$ of H_2 onto I^∞. Clearly $f \times 1$ has no extension onto I^∞.

It is yet unsettled when $n \geq 3$.

In regard to (2), the author in [14] shows:

THEOREM 4. *There exists a "wildly embedded" Cantor set* C *in* I^∞.

We remark that the construction of C is an inductive modification of Antione [4] and of Blankinship [5]. In fact at any finite stage n, the n-tube and n-chains are obtained exactly the same way as in [5], except we require that each chain must intersect I^n in a way that the intersection is a (roughly speaking) straight portion of the whole chain. We further remark that no compact subset in $\overset{\circ}{I}{}^\infty$ (or ℓ_2) can be wildly situated in $\dot{I}^\infty$ [1] [2] [9]. *Theorem 4* therefore implies:

COROLLARY 4. *There exist two Cantor sets in the Hilbert cube such that no homeomorphism of one onto the other can be extended to a homeomorphism on the Hilbert cube.*

It is interesting to know that using the technique in proving *Theorem 4*, we can also show:

COROLLARY 5. s *and* ℓ_2 *contain zero-dimensional closed sets whose complements are not simply connected.*

BIBLIOGRAPHY

[1] R. D. Anderson, *Topological Properties of the Hilbert Cube and the Infinite Product of Open Intervals,* to appear in Feb. 1967, Tran. Amer. Math. Soc.

[2] ______________, *Hilbert Space is homeomorphic to the Countable Infinite Product of Lines,* Bull. Amer. Math. Soc., 72 (1966) 515-519.

[3] ______________, (Abstract), *On Extending Homeomorphisms on the Hilbert Cube,* Notice Amer. Math. Soc., Vol. 13, No. 3 (Apr. 1966), p. 375.

[4] L. Antoine, *Sur l'homeomorphie de deux figures et de leurs voisinages,* J. Math. Pures. Appl., 86 (1921) 221-235.

[5] W. A. Blankinship, *Generalization of a Construction of Antoine,* Ann. of Math., 53 (1951) 276-297.

[6] M. Brown, *Locally Flat Imbeddings of Topological Manifolds,* Ann. of Math., 75 (1962) 331-341.

[7] ______________, *A Proof of Generalized Schoenflies Theorem,* Bull. Amer. Math. Soc., 66 (1960) 74-76.

[8] M. Brown and H. Gluck, *Stable Structures on Manifolds*: I, Ann. of Math., 79 (1964) 1-17.

[9] V. Klee, *Convex bodies and periodic homeomorphisms in Hilbert space,* Tran. Amer. Math. Soc., 74 (1953) 36.

[10] ______________, *Homogeneity of Infinite-Dimensional Parallotopes,* Ann. of Math. (2) 66 (1957) 454-460.

[11] ______________, *Some Topological Properties of Convex Sets,* Trans. Amer. Math. Soc., 78 (1955) 30-45.

[12] ______________, *A note on Topological Properties of Normed Linear Spaces,* Proc. Amer. Math. Soc., 7 (1956) 673-674.

[13] R. Wong, *A Theory of Homeomorphism on Certain Infinite Product Spaces,* (to appear in Tran. Amer. Math. Soc.).

[14] ______________, A Wild Cantor Set in the Hilbert Cube, (to appear).

[15] ______________, A proof that any homeomorphism on the Hilbert Space is isotopic to the identity mapping, (to appear).

UNIVERSITY OF CALIFORNIA, LOS ANGELES